POCKET
ANATOMY

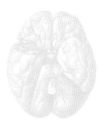

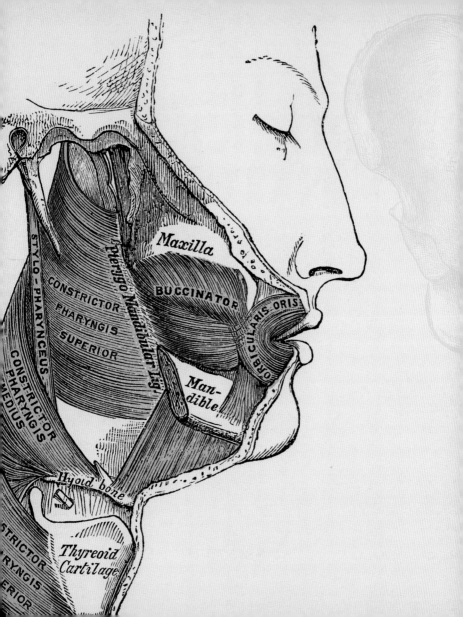

POCKET
ANATOMY

A Complete Guide to the Human
Body, for Artists and Students

Christopher Joseph

First edition for the United States, its territories and possessions, and Canada, published in 2006 by Barron's Educational Series, Inc.

Design copyright © THE IVY PRESS LIMITED, 2006
Text copyright © CHRISTOPHER JOSEPH, 2006

All inquiries should be addressed to:
Barron's Educational Series, Inc.
250 Wireless Boulevard
Hauppauge, New York 11788
www.barronseduc.com

ISBN-10: 0-7641-5908-9
ISBN-13: 978-0-7641-5908-4

Library of Congress Catalog Card No.
2005924640

This book was created by
THE IVY PRESS LTD
The Old Candlemakers, West Street,
Lewes, East Sussex BN7 2NZ, U.K.

Creative Director PETER BRIDGEWATER

Publisher JASON HOOK

Editorial Director CAROLINE EARLE

Senior Project Editor HAZEL SONGHURST

Editor STEPHANIE HORNER

Art Director SARAH HOWERD

Design CHRIS AND JANE LANAWAY

Anatomy Consultant DR. DONAL SHANAHAN

Printed in China

PUBLISHER'S NOTE: This book is intended as a general reference guide only and its contents should not be substituted for the information provided by a standard medical textbook.

CONTENTS

INTRODUCTION

Gray's Anatomy, first published in 1858, is probably one of the most famous textbooks in the English-speaking world, with its detailed, authoritative text and beautifully clear illustrations. After the early death of its original writer, successive editors ensured that it remained an accurate and complete guide to the medical knowledge of the day. By 1905 it had grown to some 1,200 pages, illustrated with around 800 drawings and diagrams showing the fine detail of every muscle, blood vessel, nerve, and other part of the body. The American edition of 1918 (from which the majority of the illustrations in this book are taken) was substantially larger, with more than 400 additional images.

For a medical student, a copy of *Gray's Anatomy* remains indispensable, and many artists who draw from life will also own a copy. However, the technical prose style remains difficult for a non-specialist to follow. For today's more general readership a different approach, retaining all the basic information while bypassing some of the more complicated medical details, is required. *Pocket Anatomy* takes the original engravings—subtle and complex works of art in their own right, combining accuracy and clarity with an elegant beauty that no modern diagram, photograph, or computer scan has ever reproduced—and combines them with a new text that is simpler, shorter, and more accessible. However, the book still explains the function of the individual components (the muscular system, digestive system, and so on) in sufficient detail to satisfy the human thirst for understanding.

Following the same basic structure as Gray, starting with the skeleton and working outward, *Pocket Anatomy* shows what each part of the human body is made from, what its name is, what it does, and how it does it. Every bone, joint, and organ is shown from different angles with respect to its various surroundings, and the way each one works is discussed, but without the complex detail and medical jargon that characterize *Gray's*. The intention with this new book is to provide as much information about the human body as possible in a way that is comprehensible to as many people as possible; and to make the text a useful aid to understanding the beautiful diagrams—some in black and white, some colored in the same meticulous way that they were drawn—rather than the other way around.

USING THIS BOOK

The illustrations for this book are taken from the 1905 and 1918 editions of *Gray's Anatomy*. Although they remain anatomically correct since the illustrations were prepared, some of the nomenclature appearing within the drawings is now outdated. In *Pocket Anatomy* the annotations accompanying the illustrations have been revised and updated in accordance with current medical terminology, and applied selectively in the interest of the general reader. In a number of the original illustrations, color was used to distinguish between the principal systems within the body, namely: red for the arteries, blue for the veins, yellow for the nerves. This color key remains unchanged.

Medical terminology has been used throughout the text and captions in this book. This includes directional terms, such as *anterior* (front or forward); common terms, such as *aural* (relating to the ear); and more specialized terms, such as *aponeurosis* (a flat tendon). A glossary on pages 308–313 explains the meanings of the medical terminology used, providing a useful source of reference for the non-specialist.

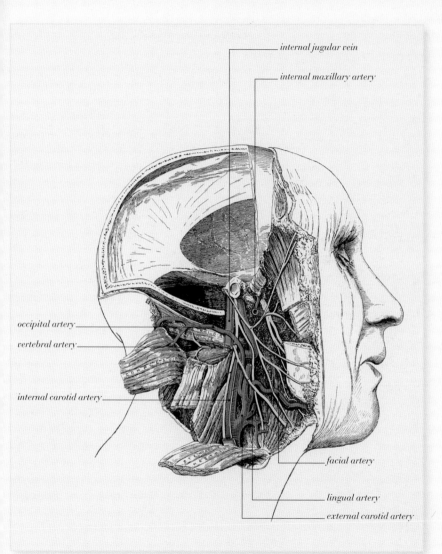

internal jugular vein

internal maxillary artery

occipital artery

vertebral artery

internal carotid artery

facial artery

lingual artery

external carotid artery

9

BONE STRUCTURE

Bone provides the main support for the human body, and its structure and properties reflect this. Bone is a natural composite, consisting mainly of calcium phosphate and collagen (a protein which is also a major component of many other types of tissue), but it also contains numerous other chemical ingredients, and several different types of cell. The overall structure is very light and has a high strength when a compressive force is applied. It is, however, brittle, and—in adults—breaks if any attempt is made to bend or stretch it. Most bones consist mainly of a honeycomb-type structure, or spongy tissue, surrounded by an outer layer of compact tissue. In a large bone the spongy tissue provides most of the volume whereas most of the weight is concentrated in the much denser compact tissue.

The long bones of the skeleton (i.e. the main bones of the limbs) have a hollow tube down the center called the medullary cavity. This is filled with yellow bone marrow, as are the spaces in the spongy bone at the ends. Bone marrow is largely responsible for the production of blood cells, but can also help to generate new bone, muscle, and ligaments. Bone, like all body tissues, is continuously renewed even after growth usually stops—in humans—in the early twenties. Renewal is achieved by several different types of cell. Most prominent of these are osteoclasts, which break down and remove old bone; osteocytes, which strengthen existing bone; and osteoblasts, which are responsible for constructing bone material. These same mechanisms also allow bone to be repaired after an injury, or strengthened by regular activity.

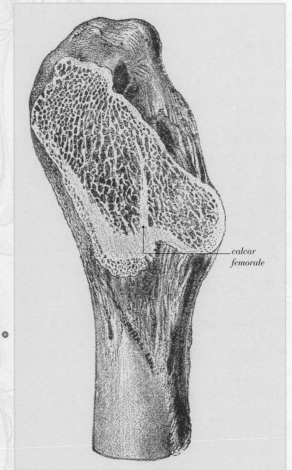

calcar femorale

Femur ○
The thighbone, cut through
to show the structure of
the spongy tissue, as well
as the much denser compact
tissue that exists around
the edge. In the femur
it also forms a reinforcing
wall (the calcar femorale)
vertically through the
middle of the main bone.

THE SKULL, FONTANELLES, AND SUTURES AT BIRTH

At birth, all the major bones of the skull have formed, but are separated from one another by various pieces of connective tissue which serve two purposes. First, during birth, they allow the skull to change shape to some extent, enabling it to pass out of the uterus and down the birth canal much more easily than if it were solid. Second, in the longer term, if the bones of the skull were fixed together at birth, it would be far more difficult for the head—and therefore the brain—to grow to any significant extent. The connective tissues are of two types: narrow fibrous ligaments (called the sutures) and larger areas of membrane (the fontanelles). The latter are the more temporary of the two; the largest fontanelle, a diamond-shaped patch close to the front of the top of the skull, becomes completely closed between one and two years after birth. The sutures are typically longer lasting. The first of these to disappear is normally the one that separates the two halves of the frontal bone. The halves may have fused almost entirely by the time a child is six or seven years old, but partial separation is not uncommon in adults and complete separation, although rare, is not unknown.

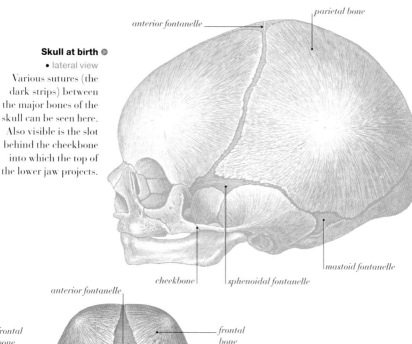

Skull at birth ○

• lateral view

Various sutures (the dark strips) between the major bones of the skull can be seen here. Also visible is the slot behind the cheekbone into which the top of the lower jaw projects.

anterior fontanelle

parietal bone

mastoid fontanelle

cheekbone

sphenoidal fontanelle

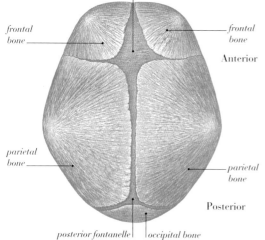

anterior fontanelle

frontal bone

frontal bone

Anterior

parietal bone

parietal bone

Posterior

posterior fontanelle

occipital bone

○ **Skull at birth** • superior view

The diamond-shaped anterior fontanelle is clearly visible; there is typically also a triangular fontanelle at the back of the head, where the two parietal bones meet the occipital bone.

THE SKULL IN GENERAL

The human skull (or cranium) is made up of a number of individual bones. The main purpose of the skull, and the sole purpose of some of the bones, is to protect the brain. However, many of the bones—particularly those in the face—are more directly involved in protecting the sense organs (eyes, nose, and tongue) and controlling the mouth's movements during chewing and speech.

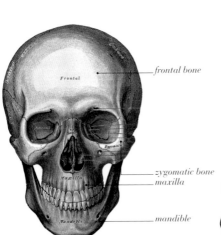

frontal bone

zygomatic bone
maxilla

mandible

● **Skull** • anterior view
One of the more unexpectedly complex areas of the skull are the eye sockets.

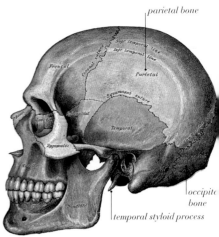

parietal bone

occipital bone

temporal styloid process

● **Skull** • lateral view
Most of the major bones and some of the smaller ones of the skull and face can be seen here. The small spike sticking forward behind the jaw is the styloid process of the temporal bone.

Base of skull • outer surface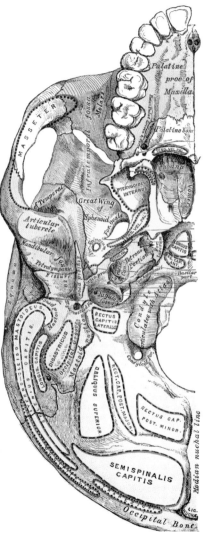

The numerous protuberances and hollows on the underside of the skull provide attachments for many of the muscles of the neck and jaw. The largest passage through the base of the skull is for the medulla oblongata, but many other nerves and several large blood vessels also have their own channels through the bone.

MAIN BONES OF THE SKULL

The neurocranium, the part of the skull containing the brain, is largely smooth with one large sunken area on each side. The underside has numerous rough patches, protrusions, and indentations, as well as a number of openings to allow the medulla oblongata, blood vessels, and nerves in and out.

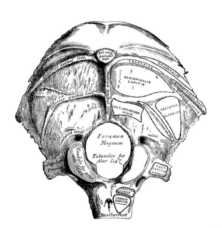

● **Occipital bone** • inferior view
This forms both the backmost part of the skull and the underside around the joint with the top of the vertebral column— the atlas *(see p. 56)*

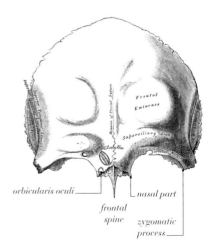

Frontal bone ●
This forms the forehead and front part of the skull, and connects to many of the bones of the face.

orbicularis oculi — *nasal part*

frontal spine   *zygomatic process*

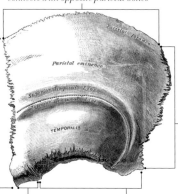

connects with opposite parietal bones

connects with occipital bone

connects with sphenoid

connects with temporal bone

connects with frontal bone

● Parietal bones

These form the main parts of the sides and top of the skull, connecting to the occipital and frontal bones at their ends and to each other along the top of the skull.

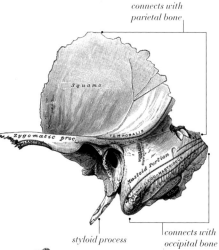

connects with parietal bone

connects with occipital bone

styloid process

Temporal bones ●

These connect to the underside of the parietal bones, and also to the occipital and sphenoid bones under the skull. In addition, they provide the point of attachment for numerous muscles around the base of the head.

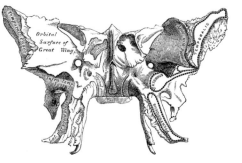

● Sphenoid bone • anterior view

This has a very complicated shape, and forms parts of the base of the cranium, the eye sockets, and the nasal cavity.

THE MAXILLAE AND CONNECTED BONES

The maxillae, sometimes called the superior maxillary bones, are two bones attached to the bottom edge of the frontal bone of the skull and to the front of the ethmoid bone, which separates the nasal cavity from the brain. As such, they form the upper jaw and part of the bottom of the eye socket. In addition, the maxillae provide the bony structures of the base and sides of the nose, and support the nasal bones (which form the bridge of the nose).

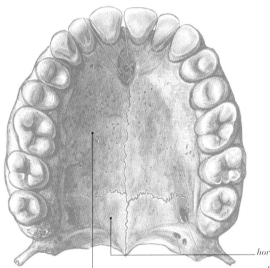

● **Palate** • inferior view

The front part of the roof of the mouth is formed by the two maxillae. The palate, farther back, is formed by a pair of bones with a roughly L-shaped cross-section. The horizontal part separates the mouth from the nasal cavity. The vertical portions of the palate bones form part of the walls of the nasal cavity and part of the eye socket.

horizontal plate of palatine bone

palatine process of maxilla

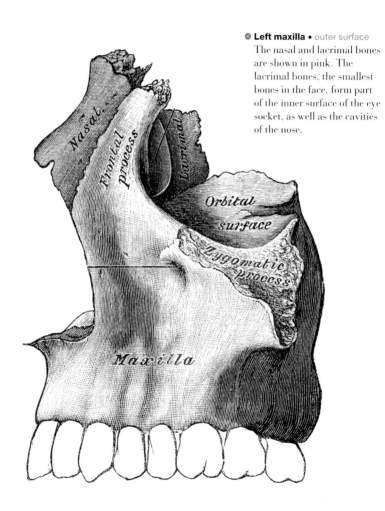

Left maxilla • outer surface
The nasal and lacrimal bones are shown in pink. The lacrimal bones, the smallest bones in the face, form part of the inner surface of the eye socket, as well as the cavities of the nose.

Nasal

Frontal process

Lacrimal

Orbital surface

Zygomatic process

Maxilla

THE ZYGOMATIC BONES

The zygomatic (cheek) bones are two of the smaller bones of the skull. They join the front edges of the temporal, frontal, and sphenoid bones and are also attached to the outside edge of the upper jaw. They form the bony part of the cheek, which projects forward from the skull, a large part of the bottom of the eye socket, and from there the entire side of the face up to and around the base of the eyebrow. Their main purpose is to support and protect the eyeball.

Zygomatic bone ◉
• shown in pink
The nasal bone can also be seen, forming the bridge of the nose. The sides and bottom of the nasal cavity are formed by the maxillae, the bones of the upper jaw.

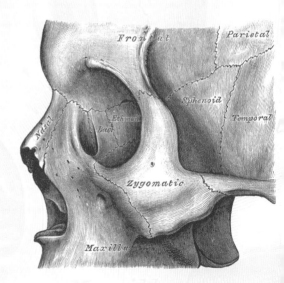

Skull • lateral view with part of the cheek and temporal bones removed This clearly shows the large socket behind the cheekbone, through which various nerves and blood vessels pass into the infratemporal fossa. Also visible is the styloid process.

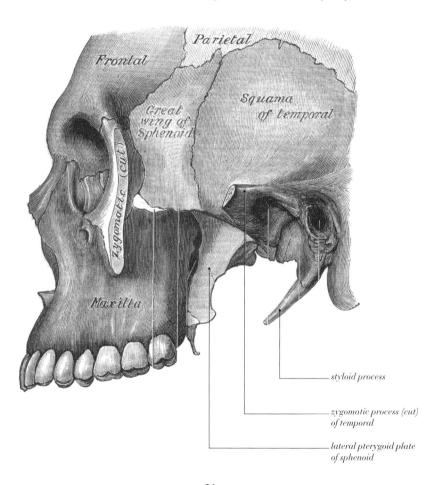

Parietal

Frontal

Great
wing of
Sphenoid

Squama
of temporal

Zygomatic (cut)

Maxilla

styloid process

zygomatic process (cut)
of temporal

lateral pterygoid plate
of sphenoid

THE MANDIBLE

The mandible (lower jaw) is a single bone, which is larger and stronger than the other facial bones. It consists of a horizontal U-shape with large vertical extensions (rami) on either side at the back of the jaw. From a medical point of view, its purpose is to support the lower set of teeth. Unlike the upper jaw, it is capable of movement. For this reason, each of the two rami divide to form a Y-shape, with the posterior section (the condyle see p. 54) articulating with the temporal bone of the skull while the anterior part (the coronoid process) connects to the temporal muscle, allowing the jaw to rotate around the condyle as this and the various muscles attached to each ramus contract and relax. The mandible is not tied to any of the other bones of the head by sutures.

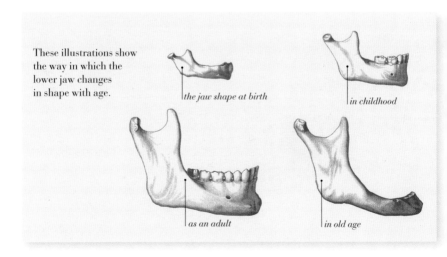

These illustrations show the way in which the lower jaw changes in shape with age.

the jaw shape at birth

in childhood

as an adult

in old age

Mandible

- outer surface, lateral view

In addition to being strong itself, the lower jaw is connected to a number of muscles, including the large temporal and masseter, in order to apply the required force when chewing.

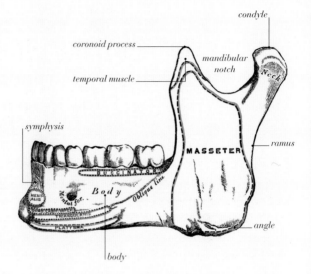

condyle

coronoid process

mandibular notch

temporal muscle

Neck

symphysis

MASSETER

ramus

BUCCINATOR

Body

Oblique line

MENT ALIS

Mental for.

PLATYSMA

angle

body

---

articulates with temporal bone

mandibular foramen

PTERYGOIDEUS EXTERNUS

TEMPORALIS

ramus

Mandibular for.

PTERYGOIDEUS INTERNUS

MYLO-HYOIDEUS

Mylo-hyoid Groove

Tongue

sup. and lingual

Mental Spine

Fossa for Submaxill. gland.

DIGASTRICUS

fossa for submandibular gland

Mandible

- inner surface, lateral view

This shows the mandibular foramen, an opening on the surface of the ramus through which blood vessels and nerves pass. Although not shown in this view, there is a horseshoe-shaped bone supporting the tongue called the hyoid bone which is suspended from the mandible.

THE VERTEBRAL COLUMN

The vertebral column, like the skull, is not a single bone but several bones working together. Since, unlike the skull, the vertebral column must remain flexible throughout life, its bones do not become unmovably attached to one another (except at the lower end), instead being permanently separated by flat discs of cartilage. While the skull has only a single main purpose—to protect the brain—the vertebral column has two: to protect the spinal cord and provide the main support for the weight of the entire upper body and head. The individual bones of the vertebral column are called the vertebrae, and are divided into groups—from the top: cervical, thoracic, lumbar, and sacral. This last section is the only part of the vertebral column that will normally fuse into larger blocks, with the sacrum and coccyx (p. 30) forming from the vertebrae below the top of the pelvis.

Vertebral column ○

• lateral view

Contrary to what might instinctively be expected, the smooth side of the vertebral column is that inside the body; the projections to the sides, some of which attach to the ribs, thrust laterally from the vertebral column itself.

The central row of projections can be felt as hard bony lumps marking the vertebral column down the middle of the back. You can also clearly see from this image the various curves of both the main vertebral column, and the sacrum and coccyx.

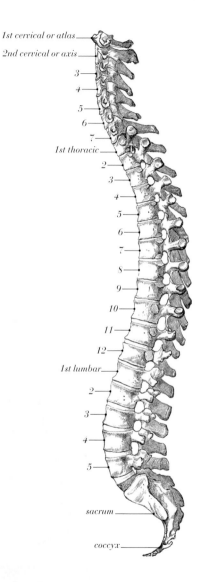

1st cervical or atlas
2nd cervical or axis
3
4
5
6
7
1st thoracic
2
3
4
5
6
7
8
9
10
11
12
1st lumbar
2
3
4
5
sacrum
coccyx

25

THE CERVICAL VERTEBRAE

The main vertebrae of the vertebral column are divided into three groups, with the cervical group at the top. Although all vertebrae have common characteristics, their appearances change significantly between the groups but are fairly similar within each one. At the top of the neck, the first two cervical vertebrae are, however, very different even from the other cervical vertebrae, but still follow the same basic pattern as the rest of their group. They are the atlas, on which the head rests, and the axis, around which it rotates.

A typical cervical vertebrae ◉

• superior view

Most of the load carried by one vertebra is passed down to the next through the vertebral body—the largest single part of the bone which remains basically stationary in any natural movement. The transverse processes, which each have a hole to allow the passage of blood vessels, are small and relatively unimportant on cervical vertebrae. The spinous process grows rapidly in size from the first cervical vertebra to the seventh, where it is extremely prominent.

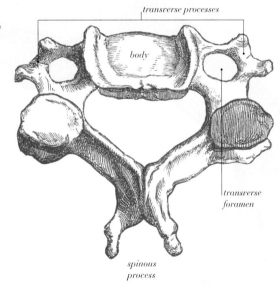

transverse processes

body

transverse foramen

spinous process

The atlas (first cervical vertebra)

• superior view

The most obvious difference between the atlas and other vertebrae is that the atlas has no body. The skull, directly above it, is able to rest on the wings at the sides, leaving a much larger space for the odontoid process of the axis and the top of the spinal cord.

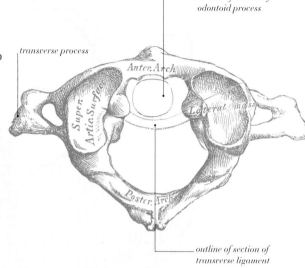

outline of section of odontoid process

transverse process

outline of section of transverse ligament

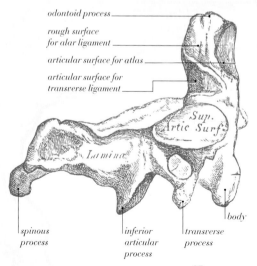

odontoid process

rough surface for alar ligament

articular surface for atlas

articular surface for transverse ligament

spinous process

inferior articular process

transverse process

body

The axis (second cervical vertebra)

• supralateral view, from left

The body of the axis extends upward above the other parts of the vertebra, forming the odontoid process. This is the main pivot around which most of the skull's rotation occurs, held in place by extremely strong ligaments to prevent it from damaging the spinal cord.

THE THORACIC AND LUMBAR VERTEBRAE

The arrangement of the lower vertebrae is much the same as for the cervical vertebrae. The body of each vertebra, though, becomes steadily larger inferiorly, to support the ever-increasing weight above. The shape of the vertebral body also changes, from the squared-off form of the cervical vertebrae, through a heart-shaped body in the thoracic vertebrae, to a large, flattened oval toward the base of the cervical vertebrae. The shape and direction of the articular processes also changes in the different vertebrae depending on the types of movements required.

an entire facet above; a demi-facet below

a demi-facet above

an entire facet

an entire facet: no facet on transverse process which is rudimentary

an entire facet: no facet on transverse process, inferior articular process convex and turned outward

● **Various thoracic vertebrae**

• right lateral view

The most obvious distinctions of the thoracic vertebrae are the large spinous and transverse processes, and the smooth areas, called facets, on the body for their joints with the ribs. Most of the ribs connect between two vertebrae, and a typical thoracic vertebra therefore has a half or demi-facet on its upper side, and another on its lower side. The first rib, and those at the bottom, each attach directly to a single vertebra, and these vertebrae therefore have a single large facet on the side.

Lumbar vertebra

• supralateral view

The spinous process of a lumbar vertebra, although larger than those of most cervical vertebrae, is much less prominent than that of a thoracic vertebra, while the articular processes are larger.

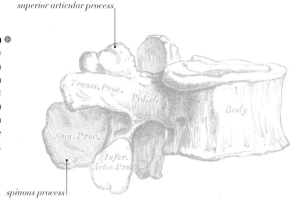

superior articular process

Transv. Proc.

Pedicle

Body

Spin. Proc.

Infer. Artic. Proc.

spinous process

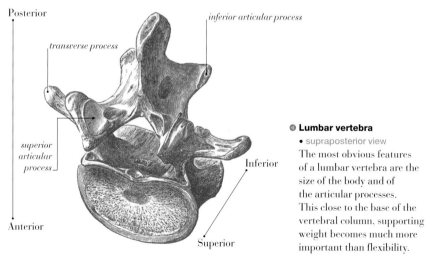

Posterior

inferior articular process

transverse process

superior articular process

Inferior

Anterior

Superior

Lumbar vertebra

• supraposterior view

The most obvious features of a lumbar vertebra are the size of the body and of the articular processes. This close to the base of the vertebral column, supporting weight becomes much more important than flexibility.

THE SACRUM AND COCCYX

The sacrum and coccyx form the base of the vertebral column, helping to transfer the weight of the upper body to the pelvis, of which the sacrum and coccyx form part. In the newborn infant they exist as groups of individual vertebrae (typically four or five for each), joining together slowly over the first 20 years or so of life to form the two main bones. It is not unknown for the two bones to fuse together as one continuous bone in mature adulthood.

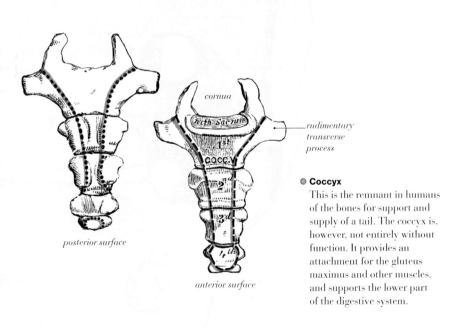

posterior surface

cornua

with Sacrum

1st COCCY.

anterior surface

rudimentary transverse process

● **Coccyx**

This is the remnant in humans of the bones for support and supply of a tail. The coccyx is, however, not entirely without function. It provides an attachment for the gluteus maximus and other muscles, and supports the lower part of the digestive system.

The sacrum connects to the innominate bones of
the pelvis on each side (*see p. 44*), to the last of
the lumbar vertebrae above it, and on the lower
side to the coccyx. The holes passing through
the sacrum on each side are called the sacral
foramina. They provide channels for nerves
emerging from the lower end of the vertebral
column and an assortment of blood vessels.

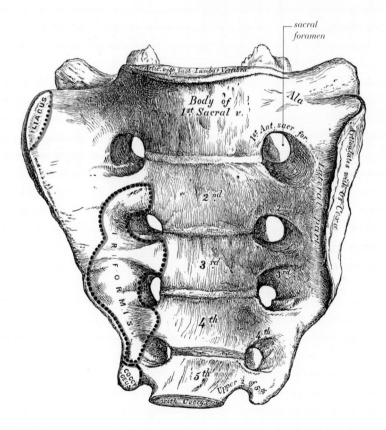

31

THE RIBS AND STERNUM

The ribs, attaching to the sternum at the front and to the vertebral column at the back, form a bony cage. Its purpose is to protect the heart and lungs, and various parts of the digestive and endocrine systems, from damage. In addition, the rib cage provides support for many muscles and connective tissues. Some of these are used to support various organs, while others are necessary for breathing—the other major function of the rib cage. The expansion and contraction of the rib cage forces air in and out of the lungs. There are normally twelve ribs on each side.

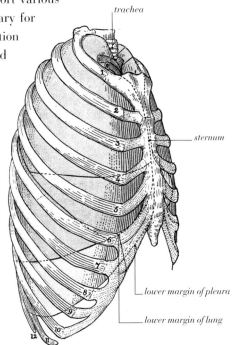

trachea

sternum

lower margin of pleura

lower margin of lung

Rib cage ○

• anterior right view

The pink area shows the position of the lungs within the rib cage: the blue area indicates the outline of the membranes protecting the lungs. The dark lines show the fissures within the lungs

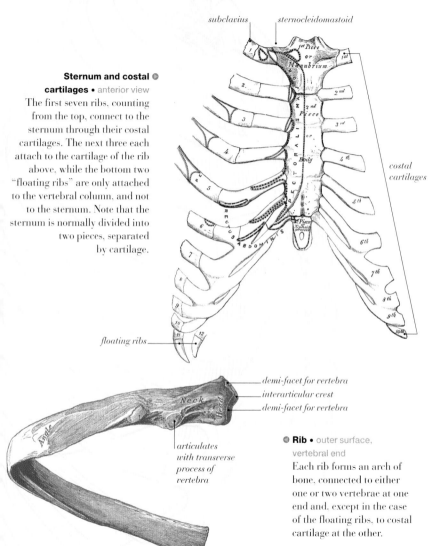

Sternum and costal cartilages • anterior view

The first seven ribs, counting from the top, connect to the sternum through their costal cartilages. The next three each attach to the cartilage of the rib above, while the bottom two "floating ribs" are only attached to the vertebral column, and not to the sternum. Note that the sternum is normally divided into two pieces, separated by cartilage.

subclavius

sternocleidomastoid

1st Piece or Manubrium

2nd Piece or Body

PECTORALIS MAJOR

RECTUS ABDOMINIS

3rd Piece Xiphoid Process

costal cartilages

floating ribs

demi-facet for vertebra
interarticular crest
demi-facet for vertebra

Neck

Head

Angle

articulates with transverse process of vertebra

● **Rib** • outer surface, vertebral end

Each rib forms an arch of bone, connected to either one or two vertebrae at one end and, except in the case of the floating ribs, to costal cartilage at the other.

THE CLAVICLE

The clavicle (or collarbone) has a gently curved S-shape. It bulges outward from its articulation with the sternum and curves back inward again as it approaches the shoulder. It serves the dual purpose of helping to support the arm and assisting the muscles of the chest, back, and upper arm in moving the limb. It has a relatively weak structure, with the shaft consisting mainly of spongy material, although the protective compact layer is thicker here than at the ends. This inherent weakness combined with its exposed position at the top of the rib cage—and the fact that it acts as a channel for much of the force of any blow applied to the shoulder—means that a fractured collarbone is the most common type of skeletal injury. The clavicle would be even more vulnerable, were it not for the fact that, as a result of its shape, it is unusually elastic for a bone.

Left clavicle • inferior surface ◗
The clavicle has attachments to the pectoral muscles of the chest, and also to the subclavius muscle, which connects the clavicle to the sternum end of the first rib. At its outward end, the clavicle articulates with the scapula *(see pp. 36–37)* to assist in moving the shoulder and arm.

● Left clavicle • superior surface

The upper surface of the clavicle connects both to the pectoral muscles of the chest and to the major muscles of the upper arm.

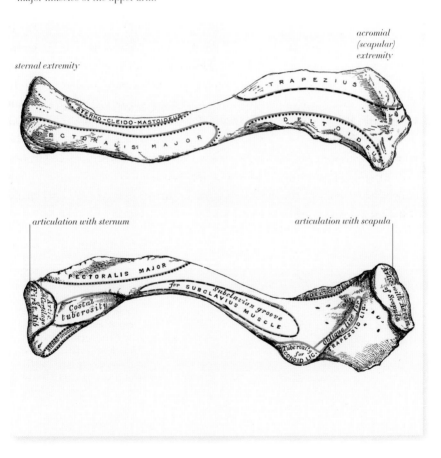

acromial (scapular) extremity

sternal extremity

articulation with sternum

articulation with scapula

THE SCAPULA

The scapula is a flat bone, roughly triangular in shape. It lies between the second and the seventh ribs, with two large protuberances from the top which head inward over the rib cage. The smaller of these (called the coracoid process) extends anteriorly, superiorly, and medially to connect at its extreme tip to the muscles of the upper arm. The larger and higher of the two (the acromion process) grows out of a broad ridge of bone across the back of the scapula and away from the body before forming a broad inward-tilting head which connects to the clavicle and to the trapezius and deltoid muscles. These two processes, along with the ridge or spine on the outside of the scapula, are all formed of spongy tissue coated with denser material. The shoulder blade itself consists solely of the denser compact bone tissue, but can be so thin in places as to be almost transparent.

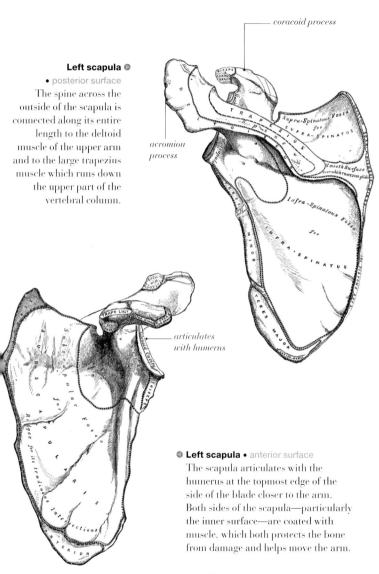

Left scapula ◉

• posterior surface

The spine across the outside of the scapula is connected along its entire length to the deltoid muscle of the upper arm and to the large trapezius muscle which runs down the upper part of the vertebral column.

coracoid process

acromion process

articulates with humerus

◉ **Left scapula** • anterior surface

The scapula articulates with the humerus at the topmost edge of the side of the blade closer to the arm. Both sides of the scapula—particularly the inner surface—are coated with muscle. which both protects the bone from damage and helps move the arm.

THE HUMERUS

The humerus (upper arm) is larger and longer than any other bone of the arm, although still substantially smaller than either the femur or tibia in the leg *(see pp. 46–47)*. The main shaft of the bone is roughly cylindrical for about half its length, becoming more triangular nearer the elbow. At the top end, it has a neck similar to that of the femur, but much shorter and wider relative to the bone. It has a smooth curved head which, unlike that of the femur, is somewhat less than a hemisphere. This allows the shoulder joint much greater freedom of movement than the hip, but as a result makes it weaker. At the lower end, the humerus connects to both the radius and the ulna *(see pp. 40–41)*—there is no equivalent of the patella *(see pp. 48–49)* at the elbow.

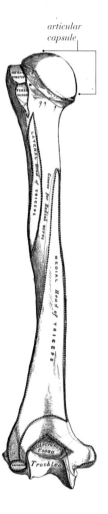

articular capsule

Left humerus • posterior view
The surface of the humerus has a somewhat twisted appearance, and is almost entirely covered by the connection points of the triceps muscle.

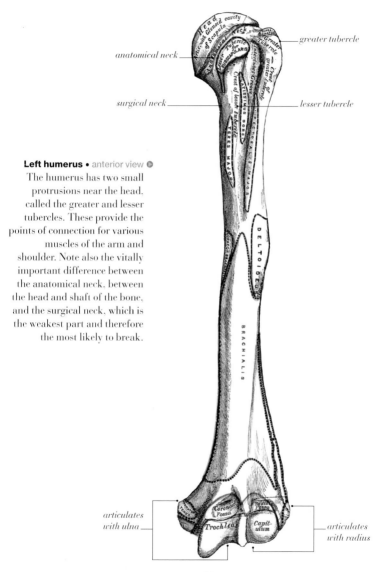

Left humerus • anterior view ◉

The humerus has two small protrusions near the head, called the greater and lesser tubercles. These provide the points of connection for various muscles of the arm and shoulder. Note also the vitally important difference between the anatomical neck, between the head and shaft of the bone, and the surgical neck, which is the weakest part and therefore the most likely to break.

anatomical neck

surgical neck

greater tubercle

lesser tubercle

articulates with ulna

articulates with radius

THE RADIUS AND ULNA

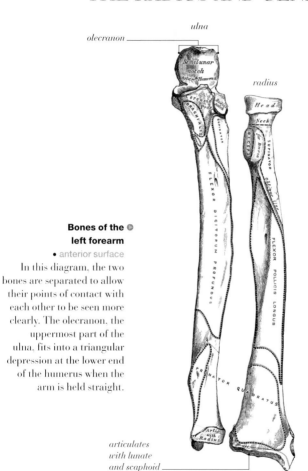

ulna

olecranon

radius

Semilunar notch Artic. w. Humerus

Head

Neck

Bones of the left forearm

• anterior surface

In this diagram, the two bones are separated to allow their points of contact with each other to be seen more clearly. The olecranon, the uppermost part of the ulna, fits into a triangular depression at the lower end of the humerus when the arm is held straight.

articulates with lunate and scaphoid

The forearm consists of two bones, both of which connect to the humerus at the elbow. Although the ulna is the longer, only the radius connects directly to the bones of the wrist and hand at its lower end; the ulna is separated from these by a thick pad of cartilage. The two bones have points of contact with one another at each end, and both are far more substantial at one end than the other. The radius covers the majority of the width of the wrist, narrowing significantly toward the elbow, while the ulna is widest at the elbow, becoming much narrower toward the hand.

**Bones of the ◐
left forearm**

• posterior surface
The round bulge of bone that can be felt on the underside of a bent elbow is the olecranon. This large process protects both the joint and the various nerves, tendons, and blood vessels that pass through it.

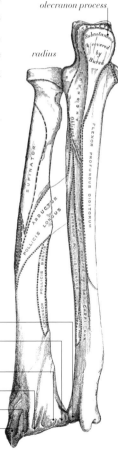

olecranon process

ulna

radius

groove for extensor carpi ulnaris

groove for extensor minimi digiti

groove for extensor indicis
groove for extensor communis digitorum

groove for extensor carpi radialis longior

groove for extensor carpi radialis brevior

groove for extensor longis pollicis

BONES OF THE HAND

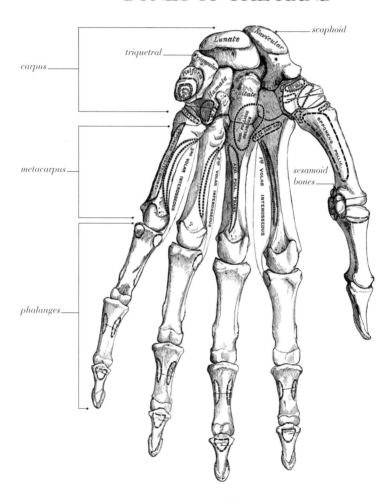

scaphoid

carpus

triquetral

metacarpus

sesamoid bones

phalanges

The bones of the hand are traditionally divided into three subgroups: the eight bones of the carpus (wrist); the five metacarpal bones, which form the palm of the hand and base of the thumb; and the phalanges, which make up the fingers. Of the carpal bones, the scaphoid and the lunate (from its slightly crescent-like cross-section) attach to the radius while the triquetral bone is separated from the ulna by the cartilage of the wrist. Of the others, the trapezium is situated at the base of the thumb, while the remaining carpal bones form the base of the palm.

Bones of the left hand
• dorsal surface
The first metacarpal bone, which forms the base of the thumb, is shorter and thicker than any of the others. However, its unusual position at an angle out to the side of the hand and the resulting freedom of movement it gains are much more important differences.

Bones of the left hand
• palmar surface
The pisiform (one of the wrist bones) forms a small hard bulge on the little finger side of the base of the palm, to which are attached two of the major tendons for moving the hand. The pisiform does not come into contact with any other bones except the triquetral. The impressive flexibility of the thumb is indicated by the large number of tendons associated with its bones.

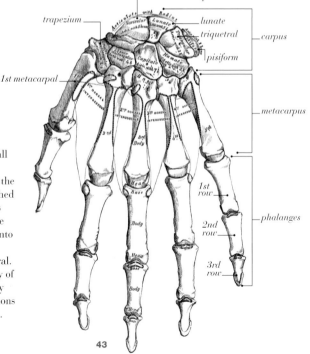

THE PELVIS AND HIP

The pelvis is a large, solid, bony ring, formed from the two innominate (nameless) bones and the sacrum and coccyx at the base of the vertebral column *(see pp. 30–31)*. The innominate bones form one side and half of the front of the pelvis. They serve a number of purposes. They support the entire body above, including the vertebral column and skull, as well as the abdomen, chest, and internal organs. Around 30 muscles are attached to each bone, including the main muscles of the abdomen, the lower vertebral column, and the upper leg. They also contain the sockets of the hip joints, through which the weight from above is transferred down into the leg and to the ground.

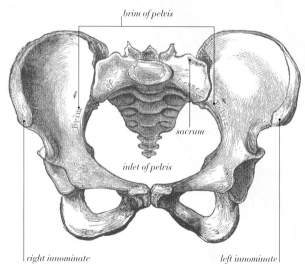

brim of pelvis

sacrum

inlet of pelvis

right innominate

left innominate

● **Female pelvis**

• superior view

The inlet of the pelvis is generally larger in women. In a man the inlet is typically around 4 inches (10 cm) from front to back, and $4^1/2$ inches (11.4 cm) from side to side, while the measurements for a woman are $4^1/2$ and $5^1/2$ inches (11.4 and 14 cm) respectively. The bones of the pelvis, however, still follow the general rule of being lighter in the female, and the depth of the "basin" is also usually less.

Right hip bone

• outer surface, infra-anterior view

The hip bone is divided into three main parts: the ilium, ischium, and pubis. The ischium, which forms the hip joint, is the lowest part of the bone and also the strongest, since it supports the other two parts and, through them, the entire body above. The pubis rises from this to the front and behind the genitals, while the ilium is a broad plate stretching to meet the inferior part of the vertebral column (the sacrum) at the back and form the crest of the hip (crest of ilium) at the side.

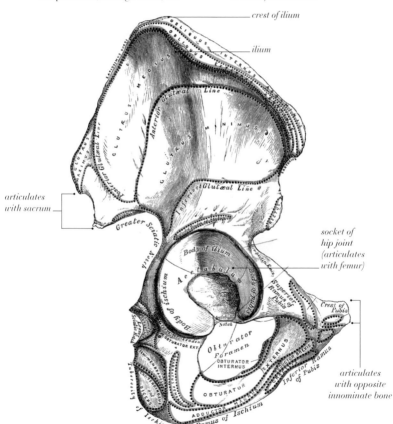

crest of ilium

ilium

articulates with sacrum

socket of hip joint (articulates with femur)

articulates with opposite innominate bone

THE FEMUR

The femur (thighbone) is the largest bone of the skeleton, in terms of both volume and length, which can be around one-third of the total height of the body. The main shaft of the femur is nearly cylindrical for most of its length, with the narrower parts formed mainly of the denser type of bone material surrounding the hollow medullary cavity. The ends have a much higher proportion of the softer spongy material surrounded—and in places interrupted—by thin compact layers. At the top of the shaft, a neck of this material, including a vertical reinforcement named the femoral spur, extends upward and sideways from the main shaft to the almost spherical head.

Right femur • posterior surface ◉
The rear of the lower end of the femur separates into two smooth protuberances called the condyles. Although these are different in size, they are actually level within the body, as the femur slopes inward from the hip toward the knee.

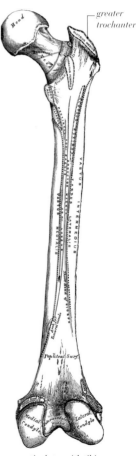

greater trochanter

articulates with tibia

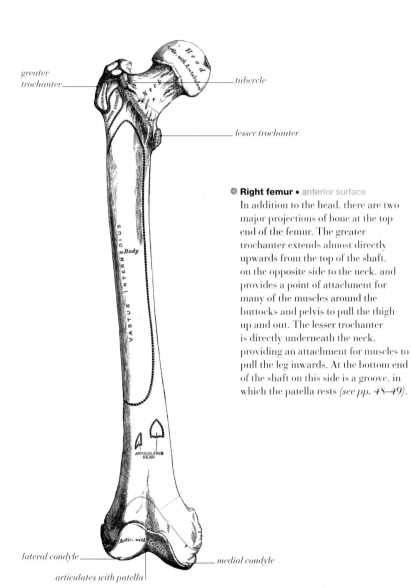

greater trochanter

tubercle

lesser trochanter

VASTUS LATERALIS

Head Artic. with Acetabulum

Neck

VASTUS INTERMEDIUS

Body

ARTICULARIS GENU

Artic. with Patella

lateral condyle

articulates with patella

medial condyle

● **Right femur** • anterior surface

In addition to the head, there are two major projections of bone at the top end of the femur. The greater trochanter extends almost directly upwards from the top of the shaft, on the opposite side to the neck, and provides a point of attachment for many of the muscles around the buttocks and pelvis to pull the thigh up and out. The lesser trochanter is directly underneath the neck, providing an attachment for muscles to pull the leg inwards. At the bottom end of the shaft on this side is a groove, in which the patella rests _(see pp. 48–49)._

THE PATELLA

The patella, or kneecap, is roughly triangular in shape, with a point at the bottom. It serves two main purposes: first, it helps to protect the potentially fragile knee from damage to the front and, second, being attached to the tendon of the large quadriceps muscle that runs down the front of the thigh, it increases the leverage available and allows the muscle to exert greater force. The fact that it forms in the tendon of the quadriceps muscle technically makes it a sesamoid bone—a bone formed in a tendon where it passes over a joint—although it is far larger than any other bone in this category.

Patella • posterior view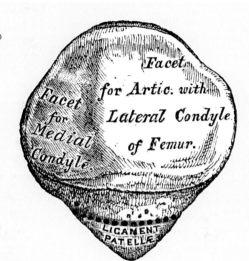

The rear of the patella consists of two almost flat surfaces covered with cartilage, separated by a vertical ridge that fits into the groove along the lower part of the femur (pp. 46–47). During movement, the kneecap slides up and down this groove as the leg bends and straightens.

Facet for Artic. with Lateral Condyle of Femur.

Facet for Medial Condyle.

LIGAMENT PATELLÆ

QUADRICEPS FEMORIS

Subcutaneous & covered by a Bursa.

LIGAMENT PATELLÆ

Patella • anterior view

The front of the patella bulges forward in a curve outward from the knee joint. It is covered by an extension of the quadriceps tendon. Much of this surface can be seen and felt through the skin.

THE FIBULA AND TIBIA

connects to femur

The lower leg, like the forearm, consists of two bones: the fibula and the tibia (shinbone). The tibia is generally the second largest individual bone in the skeleton after the femur *(see pp. 46–47)*, while the fibula, which runs down the back of the leg, is the thinnest—relative to its length—of the "long bones" making up the limbs.

tibia

fibula

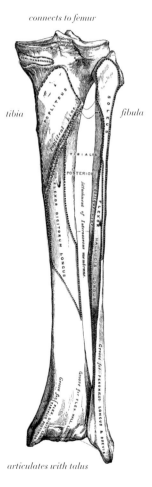

Bones of the right leg • posterior view
The fibula is behind and to the outside of the tibia: the two are connected by an interosseous membrane. The top of the fibula attaches to the back of the head of the tibia, rather than directly to the knee joint, while its lower end forms the bony bulge on the outside of the talus (ankle).

articulates with talus

Despite the fact that the tibia is almost unprotected at the front of the leg, the fibula is the more likely of the two to be broken separately, although most broken legs involve both bones rather than one or the other. In men the tibia is vertical, but in women it slants slightly outward as it descends, to compensate for the different shape of the female pelvis.

Bones of the right leg • anterior view ●
The main shaft of the tibia is roughly a triangular cross-section, broadest at the top and getting steadily narrower for about three-quarters of its length before broadening again toward the lower end. The anterior border (sometimes called the crest of the tibia) can be felt as the ridge of bone running down the front of the lower leg between the muscles.

BONES OF THE FOOT

The bones of the foot are divided into three subgroups: tarsals (heel, ankle, and rear part of foot); metatarsals (long fixed bones toward the front of the foot), and phalanges (toes). The largest of the tarsal bones is the calcaneus, which forms the heel and provides the attachment for the main muscles of the calf and the Achilles tendon. Next in size is the talus,

Bones of the right foot

• palmar (sole) view

The bones of the foot, and of the big toe in particular, are far less mobile than their counterparts in the hand, as a result of their much greater connecting service areas. The human foot is most unusual in being perpendicular to the leg—a result of the need to support our weight in an upright posture.

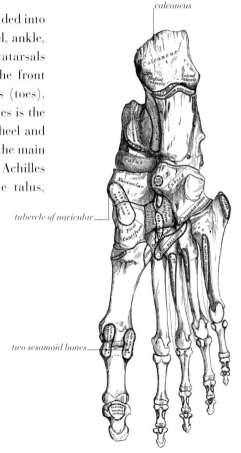

calcaneus

tubercle of navicular

two sesamoid bones

which forms the main part of the ankle, supporting the tibia and fibula *(see pp. 50–51)*. The remaining tarsal bones serve to connect the calcaneus and talus to the five metatarsals, which are numbered from the inside of the foot to the outside. Each metatarsal consists of a flattish, roughly triangular base followed by a shaft leading to a smoothly curved surface at the distal (toe) end. Beyond these are the phalanges, of which there are two for the big toe and three for each other toe, as with the thumb and fingers, respectively.

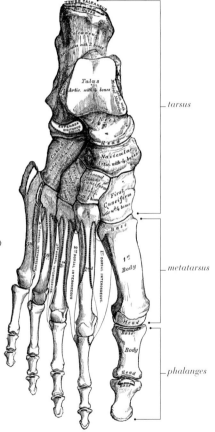

— *tarsus*

— *metatarsus*

— *phalanges*

Bones of the right foot ◉

• superior surface

The five metatarsals are numbered from the inside of the foot to the outside. The first metatarsal is both much broader and rather shorter than any of the others. The phalanges of the big toe are similarly broad, but no shorter than their companions. Note the different number of tarsal bones to which each of the metatarsals connects.

JOINT OF THE MANDIBLE

T he ligaments of the joints hold the bones securely together, allowing them to move relative to each other. The joint between the mandible (lower jaw) and the skull is relatively simple, but allows impressive freedom of movement. The jawbone can be moved forward, backward, and to a limited extent, from side to side, as well as up and down. This allows the teeth to be used in a far

Mandibular joint ◐
• outer view
The temporomandibular ligament attaches to the thin arch of the temporal bone (which connects to the outer edge of the zygomatic bone) and to the back edge of the neck of the jaw. The capsular ligament, which can be seen behind it, runs all the way round the top of the condyle of the jaw, sealing it, the synovial membranes, and cartilage disc to the temporal bone.

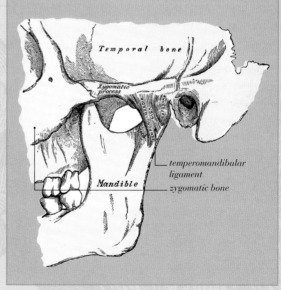

Temporal bone

Zygomatic process

temperomandibular ligament

Mandible

zygomatic bone

greater variety of chewing actions than if the jaw just hinged up and down. All the ligaments involved in the joint are connected to the condyle or to the back edge of the ramus on each side *(see p. 22)*. The coronoid process and main part of the ramus hold the attachments of the chewing muscles to the jaw, while the main horizontal U-shape is connected to muscles in the cheeks and neck.

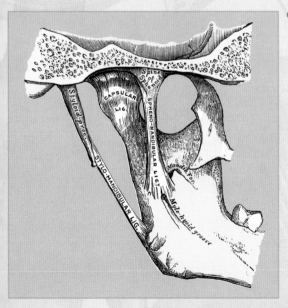

○ **Mandibular joint** • inner view
Three of the four ligaments involved in the movement of the jaw are visible here. The capsular ligament can be seen more clearly from this view. The spheno-mandibular ligament is both longer and thinner than the temporomandibular. The sphenomandibular ligament attaches to the lingula—the spur of bone overhanging the mandibular foramen. The stylomandibular ligament, although it connects the jaw and the styloid process, contributes relatively little to the movement.

CONNECTION OF THE VERTEBRAL COLUMN TO THE SKULL

The base of the occipital bone of the skull is connected by various ligaments to the first and second cervical vertebrae (the atlas and axis, respectively, see pp. 26–27). The joints of these three bones are extremely complex. This is not surprising when you consider that they allow a great part of the overall movement of the head in every direction: rotating around the spine, tilting to either side, and nodding forward and backward.

Ligaments connecting
**the occipital bone to the
atlas, and the atlas to
the axis •** anterior view
The anterior ligaments
are broad, strong sheets,
strengthened by an
upward extension of the
anterior longitudinal
ligament of the spine.

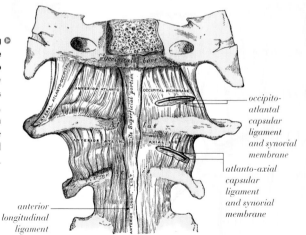

*occipito-
atlantal
capsular
ligament
and synovial
membrane*

*atlanto-axial
capsular
ligament
and synovial
membrane*

*anterior
longitudinal
ligament*

Ligaments connecting the occipital bone to the atlas, and the atlas to the axis • posterior view

Like the anterior longitudinal ligaments, the posterior longitudinal ligaments are also broad, but thinner from front to back. Those connecting the occipital bone and atlas do not attach to the bone at the sides and instead leave an open passageway used by the artery and nerve.

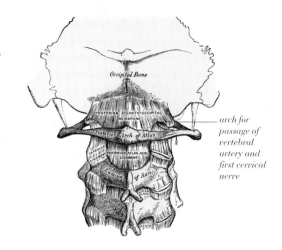

Occipital Bone

POSTERIOR ATLANTO-OCCIPITAL MEMBRANE

Posterior Arch of Atlas

POSTERIOR ATLAN-AXIAL LIGAMENT

of Axis

arch for passage of vertebral artery and first cervical nerve

occipito-axial ligament, cut horizontally and folded back

vertical portion of odontoid ligaments

occipito-atlantal capsular ligament and synovial membrane

Basilar of Occipital b

ODONTOID   ALAR LIGAM

TRANSVERSE   VERTICAL PORTION

Body

atlanto-axial capsular ligament and synovial membrane

Ligaments from occipital bone and atlas to the axis

• posterior view, with part of skull and arches of vertebrae removed

The odontoid ligaments, between the base of the skull and the atlas, are strong narrow cords which act to limit the rotation of the skull relative to the neck. The transverse ligament presses the upward extension of the axis (the odontoid process) firmly in place against the anterior of the inside of the atlas.

THE VERTEBRAE AND THEIR LIGAMENTS

The movement of each vertebra, relative to its neighbors, is limited in order to avoid damage to the nerves of the spinal cord. The combination of these movements in different directions for a large number of vertebrae, however, provides the vertebral column as a whole with an impressive degree of flexibility. The cervical region of the spine has the greatest freedom of movement, in particular at the top (see previous page). Fully mobile humans should be able to rotate their heads more than 90° relative to their shoulders, and their shoulders almost 90° relative to their hips (there are more than twice as many intervening vertebrae between shoulder and hip as between head and shoulder).

Two lumbar vertebrae ○ and their ligaments

• vertical cross-section

The supraspinal ligament is a strong cord running from the sacrum up to the lowest cervical vertebrae, although no individual fibers of the ligament connect more than three or four vertebrae. The anterior and posterior longitudinal ligaments are broader, flatter, and longer. They run from the sacrum to the second cervical vertebra, with an extension connecting to the occipital bone of the skull. The posterior longitudinal ligament, running up the inside of the vertebral column, is separated from the spinal cord by a thin sheet of loose tissue. The interspinal ligaments connect the projection of bone at the back of each vertebra to its neighbors, preventing them from moving too far apart. The flat cartilaginous discs between the body of the vertebrae themselves help to allow the vertebrae to move relative to each other, and to prevent them from being crushed together.

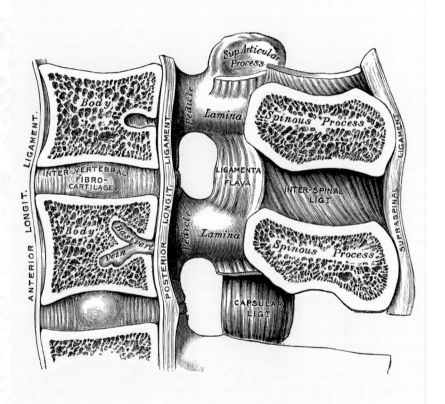

THE RIB CAGE

Although a discussion of the rib cage does not seem appropriate in a section on joints, the action of breathing moves the ribs up, down, forward, and backward, requiring them to have some flexibility in relation to the spine and—where appropriate—the sternum. At the front, the first to seventh ribs are connected to the sternum by costal cartilage, which provides a large part of the rib cage's overall movement.

Joints between the ribs and ◉ sternum • anterior view

As with most joints, a capsular ligament surrounds the connection between each costal cartilage and the sternum. In this case, the capsular ligaments are in turn protectively surrounded by the anterior chondrosternal ligaments. Unusually, however, there is no intervening synovial membrane associated with the first costal cartilage. Instead it connects directly to the sternum. The second costal cartilage has two synovial membranes and an extra ligament connected directly to the cartilage between the two parts of the sternum.

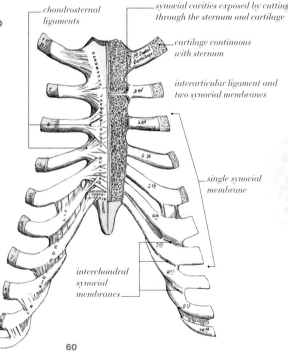

chondrosternal ligaments

synovial cavities exposed by cutting through the sternum and cartilage

cartilage continuous with sternum

interarticular ligament and two synovial membranes

single synovial membrane

interchondral synovial membranes

Joints between ribs and the vertebral column

• inner view from rib cage

The anterior costotransverse ligaments connect the upper side of each rib to the underside of the transverse process of the vertebra above. The ligaments from the head of the rib divide into three, connecting to two vertebrae and the disc between them—for those ribs that attach directly to a single vertebra, the ligaments connect to this and the one above it.

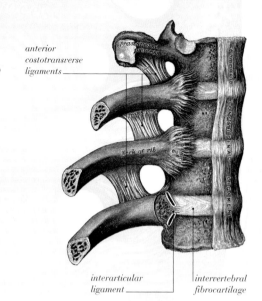

anterior costotransverse ligaments

interarticular ligament

intervertebral fibrocartilage

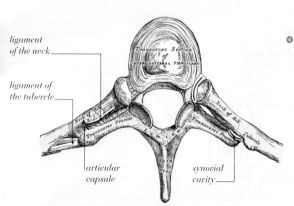

ligament of the neck

ligament of the tubercle

articular capsule

synovial cavity

Joint between a rib and a vertebra

• superior view

Most ribs connect to the spine at the joint between two vertebrae as shown here. Each pair of ribs requires a total of four synovial membranes to separate them from each pair of vertebrae and their transverse processes.

THE SHOULDER JOINT

The shoulder is necessarily complex. It is a ball-and-socket joint that combines an impressive freedom of movement with sufficient strength to allow even a relatively unfit person to support their own weight without suffering damage. The curved ball at the head of the arm is connected by ligaments to the scapula *(see p. 36)*, although these connect both to the interior of the socket in which the ball rests and to the coracoid process.

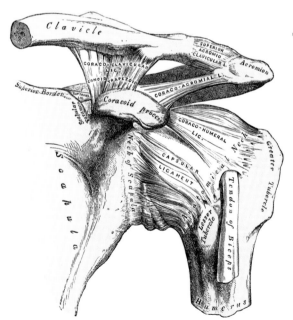

◐ **Left shoulder joint**

● anterior view

The joint between the head of the humerus and the scapula is protected by a large capsular ligament, with an opening to allow the passage of the biceps tendon. Above this—and reinforcing it at the humeral end—is the coracohumeral ligament, which connects the humerus to the coracoid process of the scapula. There are three additional glenohumeral ligaments inside the capsule, not visible here. These connect the lesser tuberosity to the glenoid cavity, the socket in the scapula into which the rounded part of the head of the humerus fits.

The scapula in turn is connected to the clavicle from both the coracoid and acromion processes and possesses ligaments connecting parts of the scapula to each other. At the other end of the clavicle, the joints to the sternum are relatively simple but provide vital leverage for the arm movements.

○ Joint between clavicle and sternum • anterior view

Like the vertebrae, the clavicle and sternum are separated by a flat articular disc. In this instance the disc is protected on each side by a synovial membrane. This arrangement in turn is protected by a capsular ligament, of which the anterior and posterior sternoclavicular ligaments form part. In addition, each clavicle is connected to the first rib immediately beneath it and to the opposite clavicle.

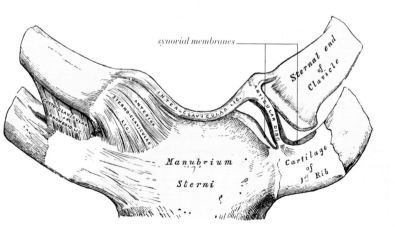

THE ELBOW JOINT

Compared with the shoulder *(see p. 62)*, the elbow is a relatively straightforward hinge joint between the humerus and the bones of the forearm. Given the enormous range of movements possible at the shoulder, the ability of the radius and ulna to rotate around each other, and the flexibility of the wrist and hand, there is no need for it to be complex. As well as being held together at the elbow, the radius and ulna are connected by several ligaments between here and the wrist. The strongest of these are the oblique ligament (which joins the ulna near the elbow and the radius much lower down) and the interosseous membrane, the fibers of which are closer to the elbow on the radius than the ulna.

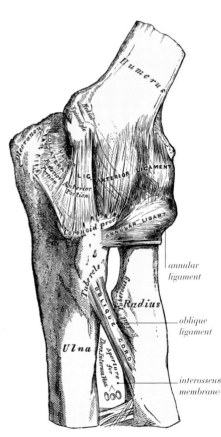

annular ligament

oblique ligament

interosseous membrane

The elbow joint is surrounded by a large capsular ligament, which varies greatly in strength. The joint is thickest at the sides, and also has thick patches front and back. These parts are sometimes referred to using distinct names for each section, but they are all fully connected.

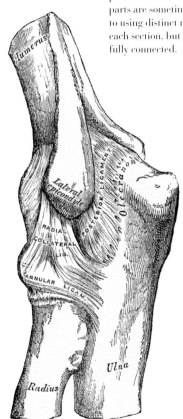

Left elbow • anterior view

The upper ligaments connecting the radius and ulna can be clearly seen here. The annular ligament surrounds the head of the radius, holding it firmly to the side of the ulna, at the elbow end. Farther down, the oblique ligament is a narrow diagonal connection between the tubercle of the ulna and the current process of the radius. The interosseous membrane is much broader (with an oval aperture for various blood vessels to pass through).

JOINTS OF THE WRIST AND HAND

Like the elbow *(see p. 64)*, the wrist joint is surrounded by a single large capsular ligament, separate parts of which are often given specific names. The wrist itself is capable of bending to a greater or lesser degree in all directions, but has no other movement. Combined with the rotation of the forearm and the flexible joints of the hand, however, the bending capacity of the wrist allows a wide range of movement. In addition to grasping with the fingers and the various movements of the thumb, the hand can be flexed inward across the palm. Along with the various tendons and muscles that control movement, the hand contains a large number of ligaments tying all the bones together.

Left wrist and ⊙
hand ligaments

• palmar (palm) view
Here the lowest of the connections between the radius and ulna can be seen, as well as the forward part of the capsule around the wrist joint. There are additional capsules around the pisiform bone, at the base of the palm, and also between the trapezium and first metacarpal bone at the base of the thumb.

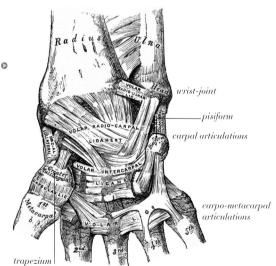

wrist-joint

pisiform

carpal articulations

carpo-metacarpal articulations

trapezium

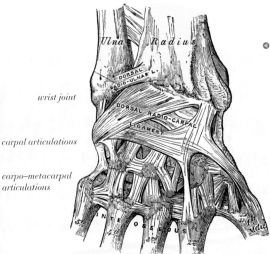

wrist joint

carpal articulations

carpo–metacarpal articulations

Left wrist and hand
- dorsal view

The dorsal ligaments on the back of the hand are generally slightly narrower than their palmar equivalents. The base of each finger is connected by lateral ligaments to the fingers on either side of it, and the subsequent finger joints each have a single cartilaginous ligament (protected by synovial membranes) on the palm side of the finger, cushioning the hinge.

Left wrist and hand synovial membranes
- palmar view

This cross-section through the bones shows the arrangement of the five synovial membranes. The third membrane is unusually complex, providing most of the internal separation between the bones of the hand.

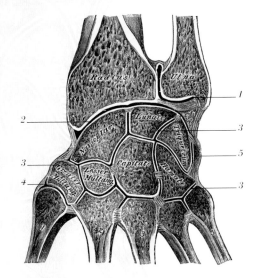

THE PELVIC JOINT

In addition to the hip joint *(see p. 70)*, the two innominate bones of the pelvis are connected to the sacrum and coccyx, to the last lumbar vertebra, and to each other by ligaments. The sacrum is joined to the last lumbar vertebra above it, in much the same way that the vertebrae further up the vertebral column are connected to each other *(see p. 28)*. It is also connected to the ilium by the iliolumbar ligament, which attaches directly to the crest of the ilium and to the sacroiliac ligament, farther down.

Pelvis (and hip) • anterior view ⊙

The anterior pubic ligament, which joins the front parts of the two pelvic bones, has a layer of diagonal fibers. These are interlaced with the aponeurosis (a flat tendon) of the external oblique muscle and the tendons at the bottom of the rectus abdominus muscles. The ligament also has a second, deeper layer of fibers that connects directly across the gap between the innominate bones. Behind these, the tips of the bones are each covered with a layer of hard cartilage, with a layer of fibrocartilage between the two. Above, below, and behind the connection are additional ligaments holding the two bones together.

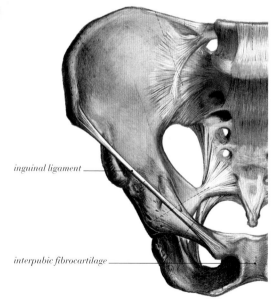

inguinal ligament ——

interpubic fibrocartilage ——

Pelvis (and hip) • posterior view

The long posterior sacroiliac ligament forms the main connection between the sacrum and each innominate bone. It consists of a large number of strong fibers and connects to both bones in several places. Below and behind it, the great sacrospinous ligament attaches to the lower part of the ilium (on the innominate bone) then—in turn—to the sacrum, to the joint between the sacrum and the coccyx, and finally back to the innominate bone at its base.

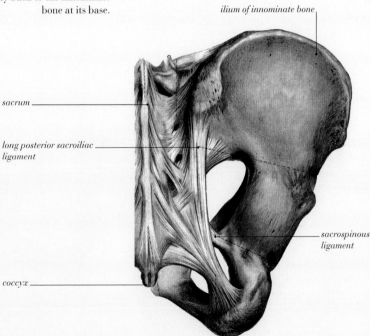

ilium of innominate bone

sacrum

long posterior sacroiliac ligament

sacrospinous ligament

coccyx

THE HIP JOINT

Paradoxically, the hip is one of the most flexible and one of the most constricted joints of the body. The main movement of the hip is rotation in any direction, albeit within restricted angles. The hip rotates around the nearly spherical head of the femur inside its socket in the innominate bone. Given the relatively simple nature of the joint, it is unsurprising that relatively few ligaments are involved. Since between them the two hips must bear the weight of the body, it is also not surprising that these ligaments are extremely strong.

Right hip • anterior view with ◯
part of the socket removed
This shows the layout of
the capsular, acetabular
labrum, and transverse
ligaments. The ligamentum
teres is triangular in shape
and is usually surrounded
by a sheath of synovial
membrane. It is connected at
its thick end to the inside of
the hip socket, and at the
narrower end to the head of
the femur.

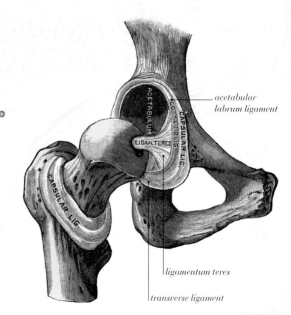

acetabular labrum ligament

ligamentum teres

transverse ligament

ilium

Iliopectin emin.a

Iliofemoral ligament

Pubo capsular Ligament

femur

Body of the femur

The large, strong iliofemoral
ligament at the front of the
hip joint connects the front
of the ilium to the femur.
It reinforces the capsular
ligament on this side as it
passes, often dividing into
two (as shown here), before
attaching to the femur.

Hip joint • cross-section ◐
This shows the arrangement
of the strong capsular and
acetabular labrum ligaments.
The latter effectively extends
the surface of the hip socket
outward to surround more of
the head of the femur. The
transverse ligament covers a
gap in the acetabular
labrum. Blood vessels run
deep to the transverse
ligament to supply part of
the femoral head.

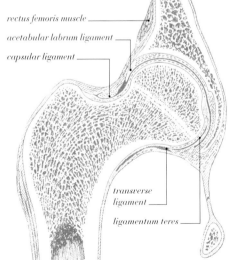

rectus femoris muscle

acetabular labrum ligament

capsular ligament

transverse
ligament

ligamentum teres

THE KNEE JOINT

Contrary to expectations, the knee is not a simple hinge joint similar to the elbow. The femur and tibia do form something close to such a joint, although their movements relative to each other are not quite those of a true hinge. Meanwhile, the patella, which slides up and down the femur protectively in front of the joint, forms, in effect, a separate joint. The fibula does not connect to the femur at all, attaching instead to a point below and behind the top of the tibia. It is not directly involved in the movement of the knee.

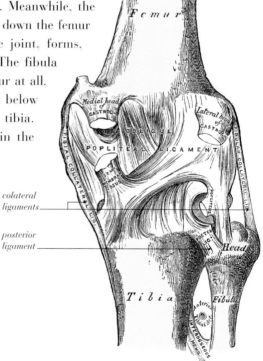

Right knee • posterior view ○
The long and short external lateral ligaments are both attached between the femur and the fibula, which thus indirectly exerts some influence on the joint. The posterior ligament is a strong band joining the head of the tibia to the opposite side of the femur, where it meets the capsular ligament (see anterior view).

colateral ligaments

posterior ligament

Right knee • anterior view ▸

The patella sits inside the middle third of the tendon of the main extensor muscles (the quadriceps) of the thigh.

This is then extended on downward from the patella to the tibial tuberosity as the patella ligament. It provides substantial additional leverage to the joint. The other two-thirds of the quadriceps tendon extend down to either side of the patella, joining the capsular ligament.

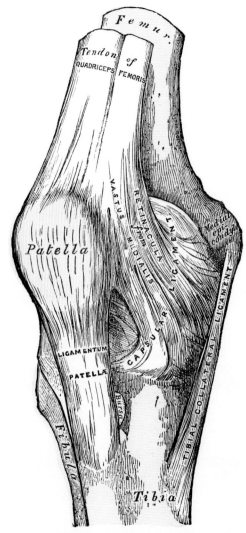

HEAD OF THE TIBIA

Strictly speaking, the head of the tibia is part of the knee joint, but it is complex enough to be worth discussing separately. The head of the tibia is surrounded by an extremely intricate arrangement of cartilage and ligaments. Unlike many joints, where separate synovial membranes are to be found between each pair of adjacent bones, the knee has only one. This starts behind the top of the patella, runs down behind the kneecap, and then follows the capsular ligament until it reaches the semilunar cartilages at the top of the tibia. It then runs along the top of the cartilages, round the back edge, and between the lower edge of the cartilages and the bone of the tibia.

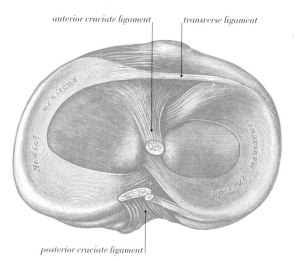

anterior cruciate ligament *transverse ligament*

posterior cruciate ligament

◉ Head of right tibia

• superior view

There are two semilunar cartilages, one on each side of the tibia. Their function is similar to the acetabular labrum ligament of the hip— extending the area of the top of the tibia available for articulation with the condyles at the bottom of the femur. The forward surfaces of these cartilages connect to the anterior cruciate ligament, while at the back they are free so that the synovial membrane can pass around them.

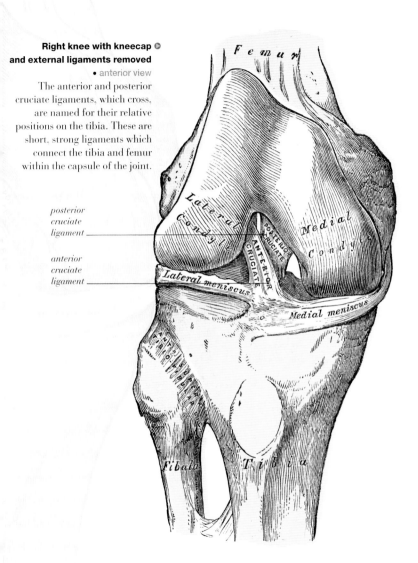

**Right knee with kneecap
and external ligaments removed**

• anterior view

The anterior and posterior
cruciate ligaments, which cross,
are named for their relative
positions on the tibia. These are
short, strong ligaments which
connect the tibia and femur
within the capsule of the joint.

*posterior
cruciate
ligament*

*anterior
cruciate
ligament*

Femur

*Lateral
Condyl*

*Medial
Condyl*

POSTERIOR
CRUCIATE

ANTERIOR
CRUCIATE

Lateral meniscus

Medial meniscus

ANTERIOR SUPERIOR

Fibula

Tibia

JOINTS OF THE ANKLE AND FOOT

The ankle is a fairly straightforward hinge joint, with a capsular ligament connecting to the base of the tibia and upper surfaces of the talus and calcaneus. The synovial membrane of the joint runs between the tibia and the talus, with a short extension from one side running up between the lower parts of the tibia and fibula.

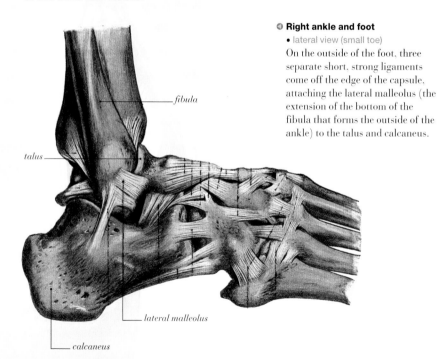

Right ankle and foot

• lateral view (small toe)

On the outside of the foot, three separate short, strong ligaments come off the edge of the capsule, attaching the lateral malleolus (the extension of the bottom of the fibula that forms the outside of the ankle) to the talus and calcaneus.

fibula

talus

lateral malleolus

calcaneus

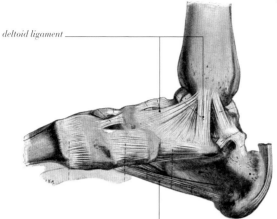

deltoid ligament

Right ankle and foot
lateral view (big toe)

The deltoid ligament is the strongest part of the capsule of the ankle. In fact, it is so strong that pulling on the ligament with ever-increasing force will usually break the bone to which it is attached before the ligament itself.

long plantar ligament

Right foot • plantar (sole) view ◉

The ligaments in the foot are arranged in much the same way as those in the hand, with short strong ligaments connecting adjacent bones both lengthwise and across the width of the foot. However, there is also the long plantar ligament, linking from the undersurface of the calcaneus, beneath the arch of the foot, to the cuboid bone and onward to the bases of the middle three metatarsals.

long plantar ligament

SOFT TISSUE TYPES

In addition to bones, blood, nerves, epithelium (the skin and linings of many internal organs), and the cells forming numerous specialized organs, the human body contains several different types of tissue. Most significant among these are connective tissue, muscle, and cartilage but there are also fat cells, medically known as adipose tissue. These store fat as a reserve of food to provide energy, and as insulation, and are found both immediately inside the skin, in a continuous layer of varying depth called the superficial fascia, and as protective coatings around various organs. Fat of various kinds is found in many other places in the body; it is also, for example, a vital and major component of both nerve cells and bone marrow.

Connective tissue comes in two main forms: a loose mesh, used to hold organs in place (relative to each other, as well as to the skin, bones, etc.); and a stronger fibrous kind. The latter can be subdivided into several different types depending on the proportions of stiff and elastic fibers. Tendons, which connect bones to muscles, and ligaments, which connect bones to other bones, are both formed mainly of stiff, white fibers—as are the fasciae and aponeuroses. Fasciae form sheaths wrapped around the muscles. Aponeuroses are thin flat tendons. They form the deep fascia, a thin layer between the superficial fascia and the muscles, but are also found elsewhere in the body. Connective tissues are repaired much less readily than most other types, and a damaged tendon or ligament may never recover fully from injury.

Cartilage, in an adult human, forms not only the joints between the bones, but also the external parts of the ear and nose and the lining of various tubes within the body. It also provides a strong junction between some muscles and the bones to which they attach. There are several forms of cartilage that contain varying proportions of fibers and cells, equivalent to the osteoblasts of bone, held together by a complex protein matrix.

Muscles are formed from bundles of fibers, held together by connective tissue, and each bundle is surrounded by a protective sheath. There are three different kinds: cardiac muscle, which is only found in the heart; smooth muscle, which is responsible for involuntary movements such as those that move food through the digestive system; and skeletal muscles, which are attached to the bone and perform voluntary movements. All three types apply force by contracting and are controlled by signals from the nervous system.

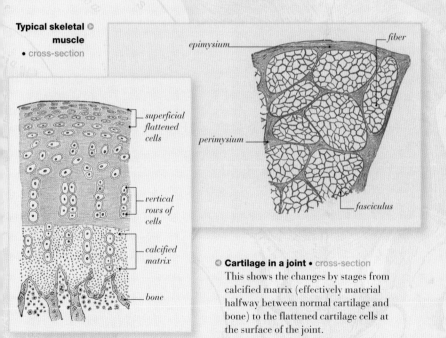

Typical skeletal muscle
• cross-section

- superficial flattened cells
- vertical rows of cells
- calcified matrix
- bone

- epimysium
- fiber
- perimysium
- fasciculus

Cartilage in a joint • cross-section
This shows the changes by stages from calcified matrix (effectively material halfway between normal cartilage and bone) to the flattened cartilage cells at the surface of the joint.

SUPERFICIAL MUSCLES OF THE FACE

The scalp has only a single, thin layer of muscle from the forehead to the base of the skull where it connects to the neck. The front part of this is responsible for raising the eyebrows and wrinkling the forehead. It connects to the muscles of the eyebrows and the nose in the face. From the top of the forehead back, it forms an aponeurosis covering the whole scalp before reverting to a thin layer of muscle about an inch above the base of the skull of the back, which connects to the large trapezius muscle of the neck. The three small muscles connected to the outside of the temporal fascia (the auricularis posterior, auricularis anterior, and auricularis superior) control the movements of the outer ear. In humans, these are unimportant, but many animals have much larger equivalents, which allow significant movements of the ear.

Surface muscles of ◎
the head and neck

• lateral view

Of the muscles attached to the nose, the procerus pulls down the inner part of the eyebrows, while the various small muscles surrounding the tip of the nose are responsible for changing the shape of the nostrils to suit different types of breathing. The long thin levator labii superioris, which runs down the side of the nose, is the main muscle responsible for dilating the nostrils (for example, when breathing hard), and also for raising the upper lip.

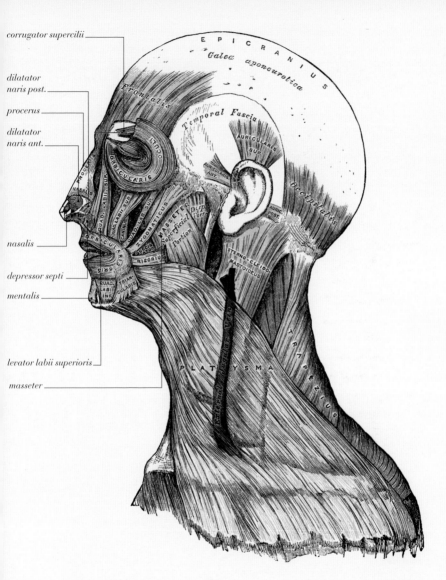

corrugator supercilii

dilatator
naris post.

procerus

dilatator
naris ant.

nasalis

depressor septi

mentalis

levator labii superioris

masseter

EPICRANIUS

Galea aponeurotica

Frontalis

Temporal Fascia

AURICULARIS
SUP.

AURICULARIS
ANT.

PROCERUS

ORBICULARIS

QUAD. LABII SUP.

QUAD. LABII SUP.

CANIN.

ZYGOMATICUS

MASSETER Deep
Superficial Portion
Portion

NASALIS

ORBICULARIS

BUCCIN.

RISORIUS

STERNO-CLEIDO-
MASTOIDEUS

AURIC.
POST.

Occipitalis

TRIAN.

QUAD.
LABII
INF.

MENTALIS

Sterno-cleido Mastoideus Ven.

PLATYSMA

TRAPEZIUS

MUSCLES OF THE EYE SOCKET

The muscles of the eye socket can be divided into two distinct groups. These are those around the outside, controlling the eyelids and the expressive area of the face around them, and those within the socket, which move the eyeball itself. Of the first group, the largest muscle is the orbicularis oculi, a sphincter entirely surrounding the outside of the socket, which is responsible for closing the eyelids. Associated with this are two additional muscles: the levator palpebrae, which raises the upper eyelid; and the corrugator supercilii, which pulls the eyebrows downward and inward when frowning. At the innermost point of the eye socket, the tensor tarsi squeezes the area around the tear ducts when these are in use.

Muscles within the right eye socket

From a narrow fibrous ring around the optic nerve at the back of the eye socket, the four recti muscles extend to four equally spaced points around the eyeball, slightly forward of its center. These are responsible for the movements of the eye, but in applying force to rotate it up or down, they also turn it slightly to one side. This is a result of the shape of the arrangement of the recti muscles. The superior and inferior oblique muscles exist to apply pressure from the other side of the eyeball, counteracting this tendency. While the superior and inferior recti of the two eyes normally work together to move both eyes up or down, looking to the left involves pulling with the lateral rectus of the left eye and the medial rectus of the right eye.

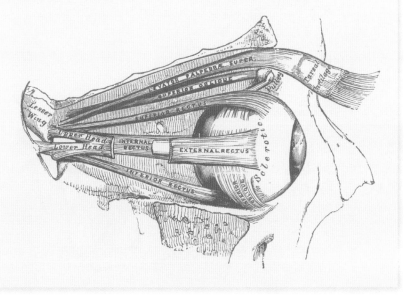

MUSCLES OF THE JAW

There are four main muscles involved in the movement of the jaw. The masseter *(see p. 81)*, temporal, and medial pterygoid muscles all connect the ramus of the jaw to different parts of the skull and act to pull the lower jaw up to meet the top jaw. The use of three large muscles, connected in this way, allows this to be done with substantially more force than could be achieved by any one muscle acting alone. The lateral pterygoid muscles, connecting the mandible and the cheek bones, control the forward movement of the jaw and can be moved independently on each side. The medial pterygoid muscles control side-to-side movement, producing a variety of grinding actions.

Pterygoid muscles ⊙

• with zygomatic arch, masseter and temporal muscles, and part of the lower jaw removed

The lateral pterygoid muscle connects to the cartilage at the top of the condyle of the ramus of the jaw. The muscle splits in two to allow the passage of the maxillary artery before connecting to two points behind the upper jaw.

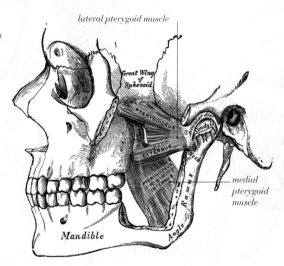

lateral pterygoid muscle

medial pterygoid muscle

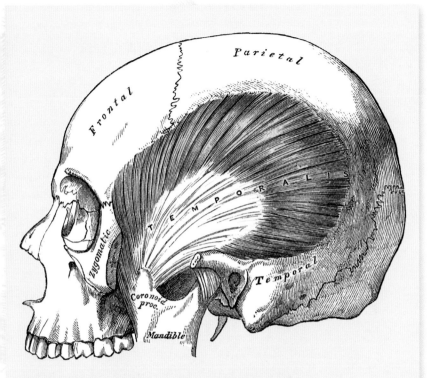

Temporal muscle • zygomatic arch and masseter removed
This muscle, which sits behind the temporal facia and masseter, consists of fibers attached to the entire inner surface of the temporal fossa, a hollow in the side of the skull. These converge into an aponeurosis, which then forms a thick flat tendon. This connects to the entire inner surface of the coronoid process at the front of the ramus, continuing almost as far as the back of the teeth.

MUSCLES OF THE TONGUE

The tongue is often said to be the only muscle in the human body that is unconnected at one end. In fact, it consists of a number of muscles, all of which connect at each end to something, although more often than not this is another of the muscles of the tongue. There are also several muscles in the space within the lower jaw connected to the hyoid bone, which assist with the movement of the lower jaw and with swallowing.

The ability of the tongue to move and change shape is vital to the production of human speech, introducing complex variations to the sound produced by the vocal cords *(see p. 254)*, as well as to the process of mastication.

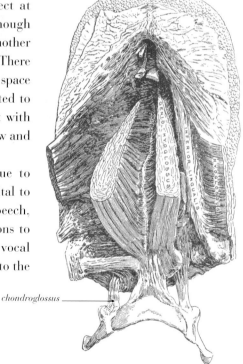

chondroglossus

Muscles of the tongue ⊙

- separated for ease of identification

Three separate muscles on either side of the tongue connect to the styloid process of the temporal bone on the same side of the skull. The styloglossus forms the side part of the tongue, while the stylohyoid raises the hyoid bone which elevates the larynx during swallowing. The stylopharyngeus also helps with swallowing by widening the top of the pharynx (see p. 88).

⊙ Muscles of the tongue

- inferior view

The main body of the tongue is made up of several different muscles on each side. The geniohyoglossus (in the middle) and hyoglossus (near the edges) pull down the parts of the tongue they form when contracted, changing the shape of the tongue. The superior and inferior linguales, at the top-center and bottom-edge, also assist in pulling down the tongue, while the styloglossus and palatoglossus raise it up relative to the jaw.

MUSCLES OF THE PHARYNX

The pharynx, at the top and back of the throat and mouth, contains muscles that are primarily involved in swallowing, although many of them are also involved in the production of speech, along with the tongue and larynx. Swallowing, like most activities we take for granted, is actually surprisingly complicated. First, the tongue moves food backward, and the entrance of the larynx is closed to prevent it from falling into the trachea and lungs. Next, the muscles of the soft palate are lifted and tensed, ensuring that the food cannot move upward into the nasal passages as it moves farther back. Muscles raise and widen the back of the throat, forming an open bag into which the food is dropped before these upper muscles relax and the constrictors contract in sequence from the top down, pushing the food slowly down into the esophagus.

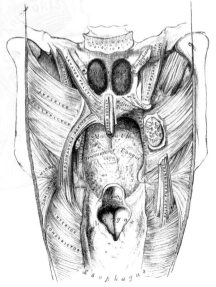

Muscles of the mouth and throat ○
• opened up posteriorly.
In addition to the constrictors (down the sides), the muscles of the soft palate can be seen at the top of this illustration. Also visible are the opening at the base of the nasal cavity, and the eustachian tubes which connect the top of the throat to the middle ear.

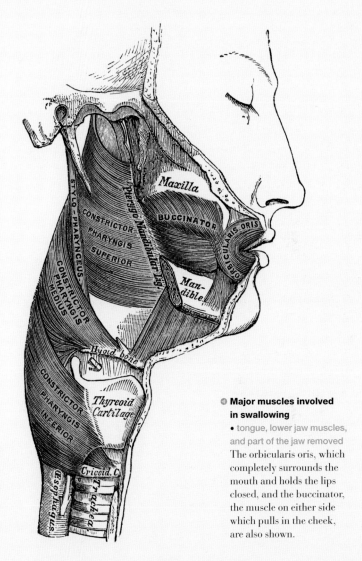

The labels on the illustration read:

Maxilla

BUCCINATOR

ORBICULARIS ORIS

STYLO – PHARYNGEUS

CONSTRICTOR PHARYNGIS SUPERIOR

Pterygo Mandibular Lig.

CONSTRICTOR PHARYNGIS MEDIUS

Man-dible

Hyoid bone

CONSTRICTOR PHARYNGIS INFERIOR

Thyreoid Cartilage

Cricoid. C.

Œsophagus

Trachea

● **Major muscles involved in swallowing**

• tongue, lower jaw muscles, and part of the jaw removed
The orbicularis oris, which completely surrounds the mouth and holds the lips closed, and the buccinator, the muscle on either side which pulls in the cheek, are also shown.

SUPERFICIAL MUSCLES OF THE NECK

The muscles of the neck carry out two separate functions: controlling the movements of the head; and providing a degree of protection for the important but fragile structures of the vertebral column, the esophagus, and the trachea which connect the head to the body. The outermost layer *(see p. 81)* is the platysma, which forms a thin tube of muscle encircling the neck. Being

Muscles of the neck ○
• lateral (left) view

The sternomastoid muscle, which runs from just behind the ear, diagonally down the neck to the sternum and inner end of the clavicle, pulls the head toward the near shoulder while rotating the face toward the opposite shoulder. The large trapezius muscle, running almost halfway down the vertebral column, is discussed with the muscles of the back *(see p. 94)*.

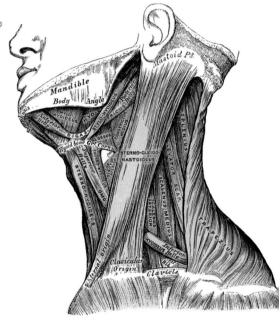

strongest at the front, contraction of the platysma pulls down the corners of the mouth. This muscle, unusually, is attached mainly to other muscles—to the pectoral and deltoid shoulder muscles at the bottom; and to the muscles and even the skin around the lower jaw and base of the skull—although some of its fibers do connect to the lower jaw.

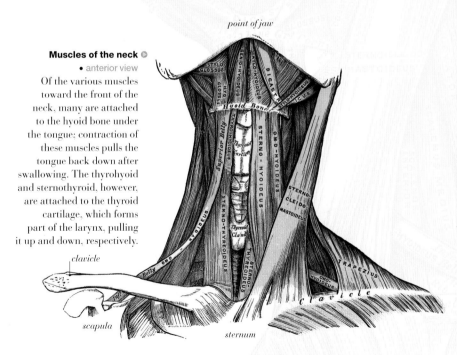

Muscles of the neck ◉

• anterior view

Of the various muscles toward the front of the neck, many are attached to the hyoid bone under the tongue; contraction of these muscles pulls the tongue back down after swallowing. The thyrohyoid and sternothyroid, however, are attached to the thyroid cartilage, which forms part of the larynx, pulling it up and down, respectively.

point of jaw

clavicle

scapula

sternum

DEEP MUSCLES OF THE NECK

The muscles within the neck can be divided into four groups. These are: those in the pharynx; those involved in the movements of the tongue, lower jaw, and larynx; the several muscles that help to raise the ribs when breathing in, and those that move the neck and head. There is considerable crossover between these last two groups. The scaleni muscles, for example, can either bend the neck sideways, bringing the ear down toward the shoulder, or raise the first and second ribs to which they are attached, depending on which end is more firmly fixed. If both muscles are contracted with the ribs fixed, the neck bends forward very slightly, with no sideways movement.

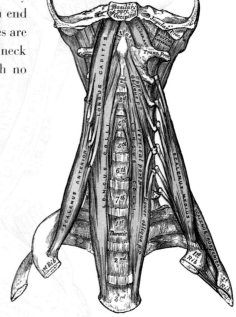

Muscles around the cervical vertebrae

• anterior view with surface muscles, esophagus, trachea, and jawbone removed

There are three scaleni muscles on either side of the vertebral column. Part of their complex attachments to the transverse processes of the cervical vertebrae can be seen. The two longus colli, which run down the front of the vertebral column from the atlas to the third thoracic vertebra, pull the head directly forward.

Muscles of the neck

• right view

The rectus capitis anterior muscles, close to the front edge of the scaleni, join them in pulling the head forward, and also turn the face to one side if only one muscle is contracted. The complexus, splenius capitis, and other muscles at the back of the neck tip the head backward, and are often also involved in rotating it.

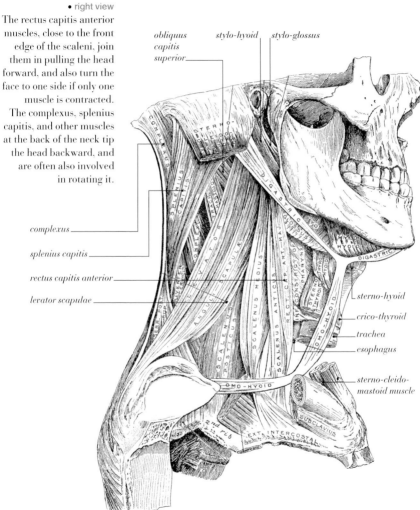

complexus

splenius capitis

rectus capitis anterior

levator scapulae

obliquus capitis superior

stylo-hyoid

stylo-glossus

sterno-hyoid

crico-thyroid

trachea

esophagus

sterno-cleido-mastoid muscle

SUPERFICIAL MUSCLES OF THE BACK

The muscles of the back, which provide support and leverage for almost everything the body does, are arranged in several layers. The outermost layer consists of two large muscles on each side: the trapezius and the latissimus dorsi at the top and bottom, respectively. Contraction of the trapezius pulls the shoulder in toward the center of the back and rotates the scapula downward. If the shoulder is held fixed by other muscles, the trapezius will pull the head back and toward it; if both are used together, the head moves directly backward. The latissimus dorsi, which connects to the lower dorsal vertebrae underneath the trapezius and forms the surface muscle over the lower back, passes round the side of the shoulder muscles to connect to the inside of the humerus, pulling it downward and inward. Within the second layer is the levator scapula, which passes down the side of the neck. It can be used either with the trapezius to pull the scapula up (for example, when shrugging) or against the trapezius to pull the neck down on that side while simultaneously rotating it. The two rhomboid muscles pull the scapula toward the vertebral column, rotating it upward. When used with the trapezius, they thus move the shoulder directly inward with no rotation. The muscles of the third layer can be divided into two straightforward groups: the serrati, which are attached to the ribs and assist with the process of breathing; and the splenii, which help to support the head and, when contracted, to draw it sideways or backward.

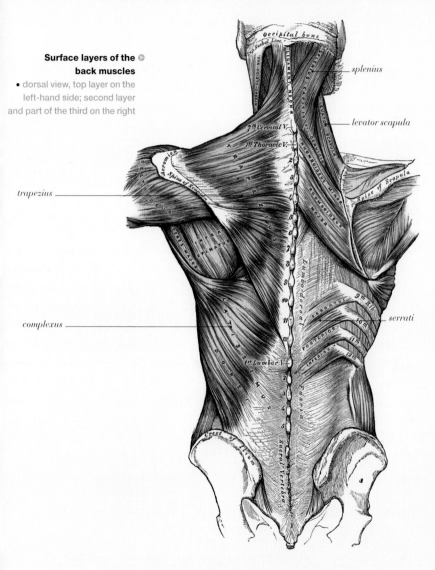

Surface layers of the back muscles

• dorsal view, top layer on the left-hand side; second layer and part of the third on the right

splenius

levator scapula

trapezius

complexus

serrati

Occipital bone

7ᵗʰ Cervical V.

1ˢᵗ Thoracic V.

Spine of Scapula

9ᵗʰ Rib

10ᵗʰ

11ᵗʰ

12ᵗʰ

1ˢᵗ Lumbar V.

Crest of Ilium

Sacral Vertebra

DEEP MUSCLES OF THE BACK

While the surface muscles of the back are largely related to movements of the rib cage, arms, and head, the deeper muscles of the back control the position and movement of the vertebral column. Both sets are heavily involved in the complex arrangements of the neck. The main purpose of the erector spinae and its constituent muscles is to keep the body upright. It can also be contracted farther than normal to pull the vertebral column backward—for example, when carrying a heavy weight or when pregnant. Its two main continuations, the ilio-costalis and longissimus dorsi, which connect it to the lower ribs and the dorsal vertebrae respectively, are themselves continued on upward to help steady the head and neck. The quadratus lumborum, inside the abdomen, runs from the bottom of the rib cage to the top of the pelvis. The sacrospinalis is the lower part of the erector spinae, connecting the sacrum to the lower vertebral column. Of the smaller muscles, the spinalis dorsi bend the main body sideways while the multifidus spinae and rotatores spinae rotate the torso around the vertebral column.

• dorsal view, fourth muscular
  layer on the right-hand side;
  fifth muscular layer on the left

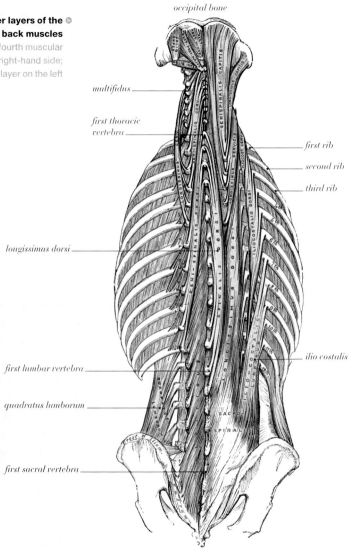

occipital bone

multifidus

*first thoracic
vertebra*

first rib

second rib

third rib

longissimus dorsi

ilio costalis

first lumbar vertebra

quadratus lumborum

first sacral vertebra

THE INTERCOSTAL MUSCLES

The intercostal (rib cage) muscles consist of two thin layers, internal and external, which run from the base of each rib and associated costal cartilage to the top of the next. The overlapping fibers, and the relatively large proportion of tendon-fibers within the muscles, give high resistance to damage, helping to protect the chest. There are layers of fascia surrounding the intercostal muscles and separating the layers, called the intercostal fascia.

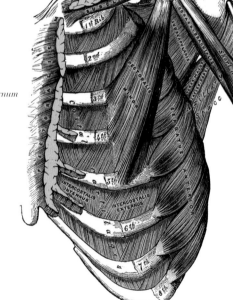

sternum

External and internal layers
of the intercostal muscles

• ventral view

The external layer
extends from the back to
the costal cartilage, and
the internal layer extends
from the sternum to
the angle of the rib.

The levatores costarum are small muscles running from each of the seventh cervical and the upper eleven thoracic vertebrae to the two ribs below each vertebra. Tensing these and the intercostal muscles lifts the ribs up and out, drawing air into the lungs. The triangularis sterni is a thin sheet of muscle and tendon-fibers on the inside of the chest cavity. It starts at the bottom third of the sternum and the costal cartilages of the lower true ribs, and runs upward and outward to the cartilages and bone of the second, third, fourth, fifth, and sixth ribs. Contracting it draws the sternum and the ribs inward, expelling air from the lungs. The exact shape and attachment of this muscle is highly variable, differing not only between individuals but often between the left and right sides of one person.

The asymmetry of the triangularis sterni is apparent: this particular example shows a slightly unusual attachment to the first rib.

sterno-mastoid

triangularis sterni

THE DIAPHRAGM

The diaphragm is a domed layer of muscle separating the abdomen from the chest cavity. The muscle fibers radiate from a central area of tendon and attach to various points around the edge of the thoracic cavity. These points include: the cartilage below the sternum; along the lower edge of the rib cage; along an arch of tendon connecting the twelfth rib to the first lumbar vertebra (the lateral arcuate ligament); to another tendon (the medial arcuate ligament), which loops across the two topmost lumbar vertebrae; and to the vertebrae themselves by medial and lateral crura. The top surface of the diaphragm is attached by a central tendon to the pericardium *(see p. 142)* and to the pleurae of the lungs at the sides; the lower surface of the diaphragm is covered by the peritoneal membrane.

When relaxed, the diaphragm is much higher in the center than at the edges. Tensing the muscle shortens the fibers, pulling the center down by up to two inches, therefore increasing the size of the chest cavity. In normal breathing, the diaphragm, intercostal muscles, and levatores costarum are used to inhale, and exhalation is caused by the rib cage falling back as these muscles relax. When gasping for breath, the muscles of the shoulder and upper back help lift the rib cage, and the abdominal muscles and the triangularis sterni force exhalation.

Diaphragm • inferior view

There are several openings in the diaphragm to allow structures to pass from the abdomen to the chest. The three largest openings are for the abdominal aorta, the vena cava, and the esophagus, but there are also several smaller ones for nerves and minor blood vessels.

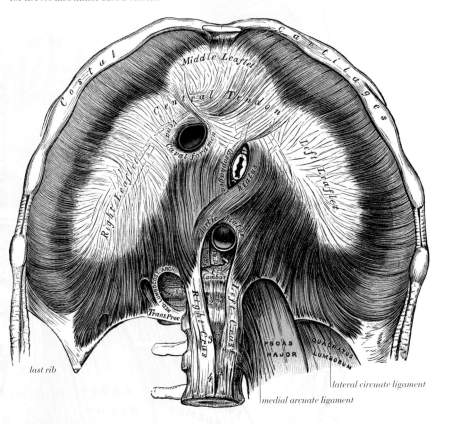

last rib

medial arcuate ligament

lateral circuate ligament

THE ABDOMINAL MUSCLES

There are three flat layers of muscle in the abdomen extending around the sides (from the outside in: the external oblique, the internal oblique, and the transversus abdominus), and a pair of vertical muscles at the front called the recti abdomini. The latter are encased in the aponeuroses of the flat muscles, which join to form lines of tendon called the semilunar lines on the outside, and the linea alba between the recti. The recti are divided into "six packs" by horizontal patches of tendon, the transverse lines. These abdominal muscles are used to apply pressure on the organs inside them—for example, when vomiting—and to flex the torso, pulling the chest down toward the pelvis.

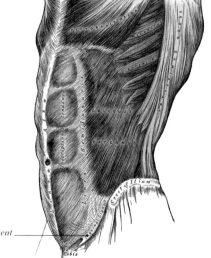

External oblique muscle ◉
This runs from the eight lower ribs vertically down to the crest of the ilium, then down and forward where it becomes an aponeurosis, which covers the whole of the front of the abdomen. The upturning of the aponeurosis forms the inguinal ligament.

inguinal ligament

Internal oblique muscle

The fibers of the internal oblique muscle radiate outward from the iliac crest and inguinal ligament. The fibers toward the front become tendinous and attach to the pubic bone, while those from the top of the crest of ilium attach to the lower four ribs. Those in the middle become an aponeurosis, which divides to sheathe the rectus abdominus muscle, rejoining at the linea alba. The lower layer is also attached to the cartilage of the seventh, eighth, and ninth ribs.

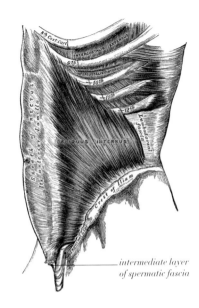

intermediate layer of spermatic fascia

tendinous intersections

linea alba

Transverse muscle

This is attached to the bottom edge of the rib cage, the lumbar fascia, the crest of ilium, and part of the inguinal ligament. Its fibers run forward almost horizontally. It terminates in an aponeurosis, which joins with the lower layer of the internal oblique aponeurosis, and then the linea alba.

MUSCLES OF THE SHOULDER AND FRONT UPPER ARM

The complexity and flexibility of the shoulder as a joint is achieved with a surprisingly small number of muscles. The two largest of these are the pectoralis major and the deltoid, which between them cover the front, top, and most of the back of the shoulder with muscle. Both muscles are roughly triangular, connecting a long sweep of bone at one end to strong tendons

Surface muscles of the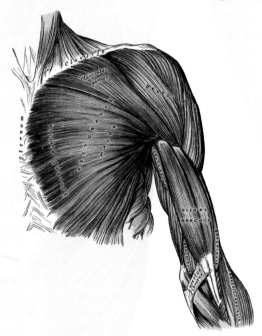
chest and front of upper arm
In addition to the sternum and half the length of the clavicle, the pectoralis major also attaches to the costal cartilages of most— and sometimes all—of the first seven ribs and even to the aponeurosis at the top of the external oblique muscle. The deltoid, the front edge of which connects to the other half of the clavicle, then runs around the edge and top surface of the acromion process and then along the spine of the scapula.

close to the head of the humerus. Underneath the main pectoral muscle is the pectoralis minor, which connects the coracoid process of the scapula to the third, fourth, and fifth ribs, and also to the aponeurosis covering the intercostal muscles. The middle fibers of the deltoid abduct the upper arm, the pectoralis major pulls it around to the front of the body, and the pectoralis minor pulls the shoulder down.

Muscles of the chest ◉
In addition to showing the pectoralis minor, this image shows the long and short heads of the biceps (the muscle that raises the forearm), which are attached to the supraglenoid tubercle and to the coracoid process respectively. The coracobrachialis, which is inside the upper arm, pulls the humerus around in front of the chest.

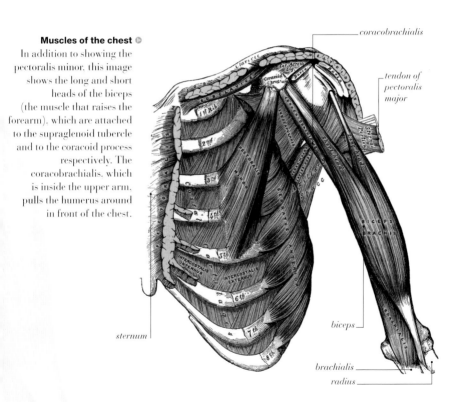

coracobrachialis

tendon of pectoralis major

biceps

brachialis

radius

sternum

MUSCLES OF THE SCAPULA AND TRICEPS

A large number of muscles connect to the scapula, allowing it to provide a great deal of leverage for the movements of the arm. In addition to the connections for the deltoid, biceps, pectoralis minor (see previous page), and trapezius *(see p. 94)* on the acromion process and spine of the scapula, several muscles are connected to the flat sides of the scapula. On the inside, against the ribs, almost the entire surface is covered by the subscapularis. The fibers of this muscle converge to form a tendon that connects to the capsular ligament of the joint and also to the lesser tuberosity of the humerus, thereby helping to prevent dislocations of the shoulder and to rotate the shoulder forward. Between the subscapularis and ribs is the serratus anterior, which is important both for pushing actions and also for helping the trapezius to abduct the arm. The large triceps muscle, which runs down the back of the humerus before connecting to the olecranon of the ulna, is used to straighten the arm.

Triceps and the muscles on outside of the scapula

Four muscles attach the back of the scapula to the humerus; the supraspinatus, infraspinatus, and teres minor all connect to the great tuberosity (greater tubercle) of the humerus, while the teres major joins the bone slightly farther down. The first three of these all assist in preventing the dislocation of the joint. The supraspinatus helps the deltoid to raise the arm, while the other three are all involved in moving the arm behind the shoulder.

MUSCLES OF THE FRONT OF FOREARM AND HAND

The forearm contains a large number of muscles, much smaller than those related to the upper arm and shoulder, but still moderately strong. Some of these provide the main power for most of the movement of the hands and wrists, while others act on the forearm— notably the pronator teres and pronator quadratus, which together rotate

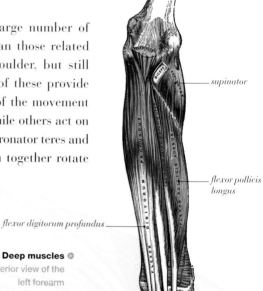

supinator

flexor pollicis longus

flexor digitorum profundus

Deep muscles ◉

• anterior view of the left forearm

The flexor digitorum profundus and flexor pollicis longus bend the joints of the fingers and thumb, respectively. Part of the supinator, the second of the muscles which rotate the radius away from the body, can also be seen just below the elbow.

the radius around the ulna toward the body. As a rule, the deeper muscles move the fingers and thumb, while those closer to the surface control the wrist and movement of the forearm. All of these muscles end in tendons or aponeuroses connecting them to the various bones of the hand, and to the wrist end of the radius and ulna.

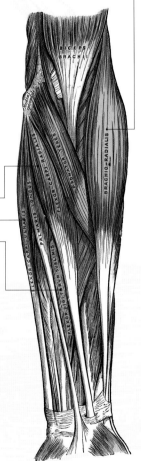

brachioradialis

flexor carpi radialis

flexor carpi ulnaris

flexor digitorum superficialis

Surface muscles • anterior view of the left forearm ◑
The flexor carpi radialis and flexor carpi ulnaris bend the wrist sideways toward their respective bones; working together, they flex the wrist (i.e. move the palm of the hand toward the underside of the forearm). If they are contracted farther, they assist with the bending of the elbow. The brachioradialis is one of two muscles that work to rotate the forearm in the opposite direction to the two pronators (i.e. to move the wrist end of the radius back away from the body). Also visible here is the flexor digitorum superficialis, which acts with the flexor digitorum profundus to bend the fingers.

MUSCLES OF THE BACK OF FOREARM AND HAND

The muscles on the posterior aspect of the forearm perform, to a large extent, the opposite actions to those on the front. Thus, the supinator rotates the radius around the ulna, away from the body. The anconeus, which connects the olecranon of the radius to the back of the humerus, assists the triceps in straightening the arm. In addition to the extensor digitorum, which straightens all four fingers, there are special muscles—the extensor indicis and extensor digiti minimi, which act to straighten only the single finger to which each is attached—the index and little fingers.

Surface muscles ◉
• posterior view of the left forearm
The tendons, and in particular the complex arrangement of the tendons of the extensor digitorum, can be clearly seen. The various extensor carpi muscles raise the wrist (i.e. move the back of the hand toward the back of the forearm) and may help to straighten the elbow.

extensis digitorum

abductor pollicis longus
extensor pollicis brevis
extensor pollicis longus
tendon of extensor indicis

extensor digiti minimi

Deep muscles • posterior
view of the left forearm

There are three muscles at the back of the forearm associated with the thumb: the extensor pollicis longus, extensor pollicis brevis, and abductor pollicis longus. These three, which pull the thumb away from the palm at the first and second joints and base respectively, combine with its local muscles and unusual arrangement of bones to give it the enormous freedom of movement that it possesses.

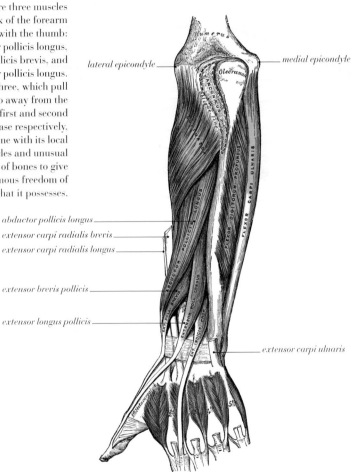

lateral epicondyle

medial epicondyle

abductor pollicis longus
extensor carpi radialis brevis
extensor carpi radialis longus

extensor brevis pollicis

extensor longus pollicis

extensor carpi ulnaris

MUSCLES OF THE THUMB

In addition to the four muscles running from the thumb up to the elbow, the three down the back of the forearm, and the flexor longus pollicis which extends up the front, there are also a number of smaller muscles within the base of the thumb and the fleshy part of the palm. Of the five muscles in the group at the base of the thumb, the adductor pollicis obliquus and adductor pollicis transversus move the thumb away from and toward the fingers within the plane of the hand. If you place your hand flat on a surface, using these muscles will neither lift your thumb off the surface nor apply pressure against it. Underneath the adductor pollicis brevis is the opponens pollicis, which moves the metacarpal bone of the thumb inward in the other sense; that is, it draws the thumb in over the palm of the hand, rotating it so as to allow gripping. The flexor pollicis brevis carries the movement farther, bending the first joint of the thumb in toward the fingers and palm of the hand.

Muscles of the left thumb

• view from above the wrist

The adductor pollicis obliquus and adductor pollicis transversus rise from a single tendon (which they share with the flexor pollicis brevis) on the side of the thumb closest to the index finger. They are attached at the far end to the capitate bone in the wrist, the base of the second metacarpal bone, and most of the length of the third metacarpal bone.

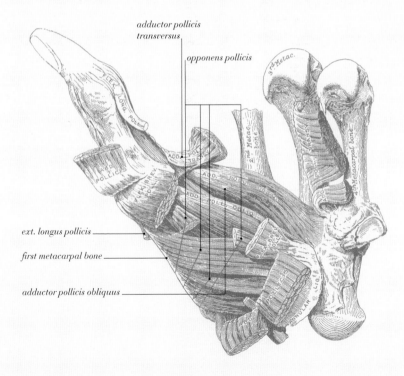

adductor pollicis
transversus

opponens pollicis

ext. longus pollicis

first metacarpal bone

adductor pollicis obliquus

113

MUSCLES OF THE PALM

Most of the movements of the hand are controlled by muscles in the arm connected by tendons to the joints of the fingers. This allows a much greater gripping strength than could be achieved by muscles within the hand, and also makes the hand less bulky than it might otherwise be.

Palm of the left hand ○

After the muscles of the thumb *(see p. 112)*, the largest muscles in the palm of the hand are usually the abductor digiti minimi, which draw the little finger out to the side of the hand, and the opponens minimi digiti, underneath it, which pull that side of the hand upward and inward, deepening the hollow of the palm. The lumbricales attach from tendons of the flexor digitorum profundus to the first joint of each finger, assisting the interossei and much larger flexor muscles of the forearm in flexing the fingers. The palmaris brevis connects the annular ligament and palmar fascia to the skin by the little finger, and helps to contract the palm sideways.

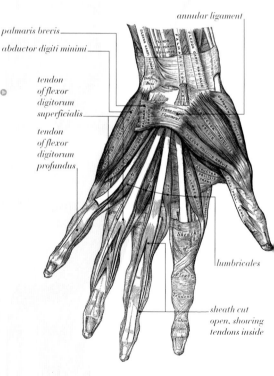

annular ligament

palmaris brevis

abductor digiti minimi

tendon of flexor digitorum superficialis

tendon of flexor digitorum profundus

lumbricales

sheath cut open, showing tendons inside

Palmar fascia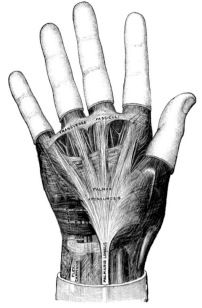

In the central part of the palm. the deep fascia. normally a very thin layer. becomes much thicker and stronger. providing a firm connection for all the various parts of the hand. It also protects the numerous nerves and blood vessels of the palm from damage.

dorsal interossei

1st   2nd   3rd   4th

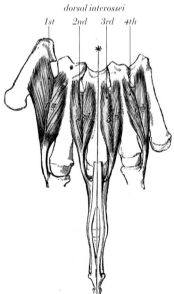

Dorsal interossei

- left hand

These muscles sit in the gaps between the metacarpal bones of the hand. attaching to the wrist end of the metacarpal bones and the base of the first phalange of each finger. The dorsal interossei. on the back of the hand, spread the fingers sideways within the plane of the hand, while the palmar interossei pull them together again.

MUSCLES OF THE PELVIS

The muscles within the pelvis divide into several distinct groups, some of which interact or act in conjunction with the various muscles that surround the pelvis. One particularly closely linked group are those controlling the anus, where the levator ani and two sphincter muscles ensure that the anal canal remains tightly shut. In the hip section, the psoas and iliacus also work together, pulling the thigh and torso together, either bending forward at the waist or raising the legs, depending on which other muscles are contracted, but many other muscles are involved in the movement of the hip joint.

Cross-section through ⊚
the pelvis • lateral view
The levator ani muscle is thin but very wide, forming a sheet that provides most of the muscular floor of the pelvis, supporting the organs within. Above and inside this is the coccygeus, which supports the coccyx. The outer sphincter muscle, which is much larger than the inner, connects to the fibers of the levator ani on each side, and also to the muscles of the perineum.

Muscles around the pelvis and hip joint ⊚
• anterior view
The quadratus lumborum, at the back, allows the angle of the pelvis to be changed. Farther forward, the psoas connects to the lumbar vertebrae and last thoracic vertebra. At the other end it forms a large tendon attached to the lesser trochanter of the femur; the iliacus, a flat muscle covering the iliac fossa on the inside of the pelvis connects to this tendon and also directly to the bone below it.

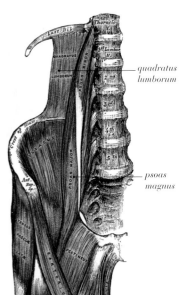

quadratus lumborum

psoas magnus

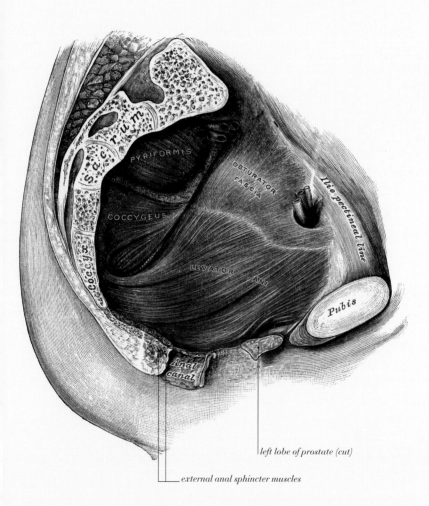

SACRUM

PYRIFORMIS

OBTURATOR
FASCIA

COCCYGEUS

Ilio pectineal line

COCCYX

LEVATOR ANI

Pubis

Anal
canal

left lobe of prostate (cut)

external anal sphincter muscles

MUSCLES OF THE PERINEUM

The perineum is the medical term for the area around the genitals: the area between the thighs and bounded by the anus at the back and the pubic bone at the front. While the muscles in this area are relatively straightforward, there are obviously important differences between the way they are arranged in men and women. The surface layer of loose connective tissue and fat cells is thicker and contains a higher proportion of fat cells here than in many other places. The deeper aponeurotic layer, although thin, is both strong in itself and firmly connected to the muscles above it. In men this provides a strong point of attachment for the muscles around the base of the penis.

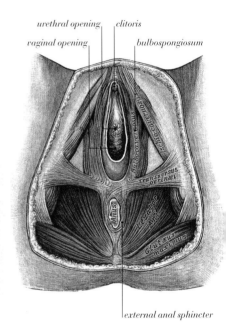

urethral opening clitoris
vaginal opening bulbospongiosum
external anal sphincter

● **Muscles of the female perineum**
 • inferior view
The erector clitoridis is the equivalent of the erector penis, while the sphincter vaginae reduces the diameter of the vagina and also compresses the vein that removes blood from the clitoris. The deep perineal fascia encloses the deep perineal muscle.

Deeper still (above the aponeurotic layer when standing upright), surrounded by the perineal muscles, is the triangular ligament (also called the deep perineal fascia). This is divided into two layers, separated at the front by an assortment of blood vessels but connected to each other at the back.

Muscles of the male perineum ◐
• inferior view
The deep and superficial transverse perineal muscles exist in both sexes to provide additional strength to the central meeting-point of the various tendons, allowing the other muscles to pull harder than would otherwise be possible.

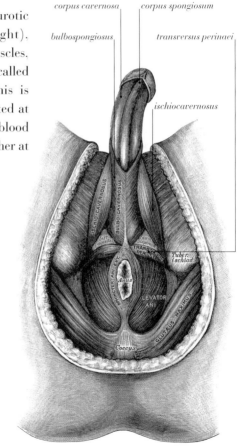

corpus cavernosa   *corpus spongiosum*

bulbospongiosus   *transrersus perinaei*

ischiocavernosus

MUSCLES OF THE HIP

The hip joint can be required to support almost the entire weight of the body at a wide range of angles. In order for this to be possible, it is necessary for both the ligaments of the joint and the muscles that move it to be extremely large and powerful. For the standard definition of strength—the ability to exert a force at the point of attachment—the strongest muscle in the body is either the gluteus maximus, which forms the outermost layer of muscle in the buttocks, or the quadriceps femoris, down the front of the thigh.

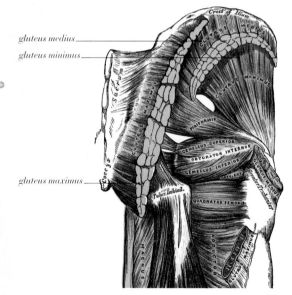

Muscles around the back of the pelvis and hip

The two inner, smaller gluteal muscles attach to the outside of the thigh, pulling it up and to the side. They are mainly used to provide assistance when standing on one leg or leaning to the side. The gluteus maximus, which surrounds the other gluteal muscles, connects to the back of the thigh at the outside edge. It is powerful enough to hold the standing body upright *(see p. 124)*.

gluteus medius

gluteus minimus

gluteus maximus

Muscles around the socket of the hip

Because the ball at the head of the femur is held tightly into the socket, the hip has very little scope for sliding movements, and every muscle that affects it must therefore do so by rotating the joint in one direction or another.

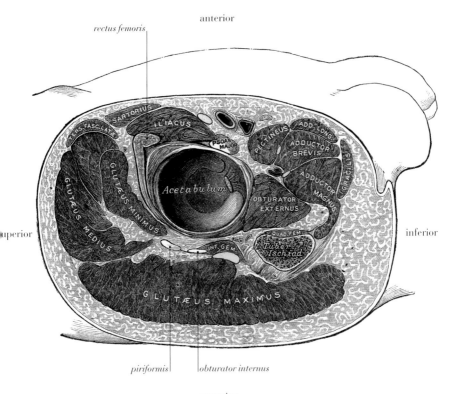

anterior

rectus femoris

SARTORIUS

TENS. FASCI. LATÆ

ILIACUS

PSOAS MAJOR

PECTINEUS

ADD. LONG.

ADDUCTOR BREVIS

GRACILIS

GLUTÆUS MINIMUS

Acetabulum

ADDUCTOR MAGNUS

OBTURATOR EXTERNUS

GLUTÆUS MEDIUS

superior

QUAD. FEM.

Tuber Ischiad.

inferior

INF. GEM.

GLUTÆUS MAXIMUS

piriformis   *obturator internus*

posterior

MUSCLES OF THE FRONT OF HIP AND THIGH

Most of the space down the front of the thigh is occupied by the quadriceps femoris, a group of four large muscles—vastus lateralis, rectus femoris, vastus intermedius, and vastus medialis—that work together to hold the leg straight. Although the quadriceps is sometimes considered a single muscle, the rectus femoris is normally separated from the others by a layer of fascia. The quadriceps femoris is connected to a variety of places at the top, but the fibers of all its sections join together in the lower part of the thigh to form a single, extremely large and strong tendon, engulfing the top of the patella. The sartorius, close to it, is the longest single muscle in the body.

Deep muscles joining the inside of ◉ the thigh to the base of the pelvis

● anterior view

The pectineus (the base end of which connects to the femur in front of the adductor brevis) and the three adductor muscles pull the thigh inward, and also rotate the knees slightly outward. Contracting these muscles provides the grip necessary to ride a horse, and is also involved in crossing the legs. Their main function, however, is to pull the rear leg forward when walking or running.

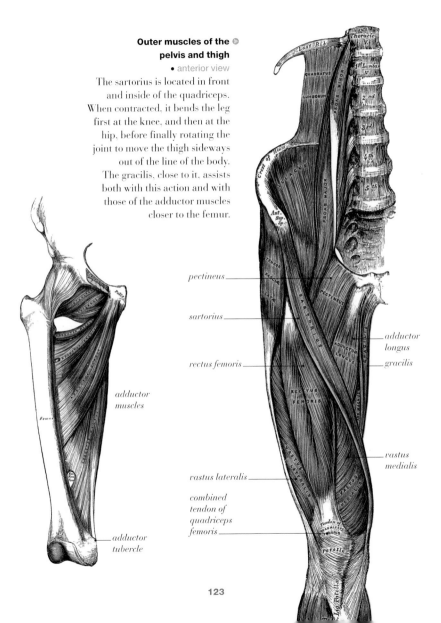

Outer muscles of the pelvis and thigh

- anterior view

The sartorius is located in front and inside of the quadriceps. When contracted, it bends the leg first at the knee, and then at the hip, before finally rotating the joint to move the thigh sideways out of the line of the body. The gracilis, close to it, assists both with this action and with those of the adductor muscles closer to the femur.

pectineus

sartorius

rectus femoris

adductor muscles

rastus lateralis

combined tendon of quadriceps femoris

adductor tubercle

adductor longus

gracilis

rastus medialis

MUSCLES OF THE BACK OF HIP AND THIGH

The muscles at the back of the thigh, and especially around the buttocks, are among the largest and most powerful to be found anywhere in the body. Like those at the front they are involved in all the movements of walking, but they are also the muscles that must be contracted in order for human beings to stand upright. This is why the hamstrings—the tendons running from the muscles of the back of the thigh down past the knee to the top of the tibia and fibula—are so important.

Muscles of the hip, buttocks, ◉
and back of the thigh

● posterior view

The annotations show the position of each muscle's tendon as it passes the knee. The three main muscles of the back of the thigh (the biceps femoris, semitendinosus, and semimembranosus) can be used either to bend the knee or to straighten the hip joint, depending on the position and which other muscles are used with them. They also, if the semitendinous and semimembranous muscles are used without the biceps, or vice versa, rotate the lower leg slightly around the knee toward the bone to which the muscles in use connect. At the top of the leg, the various muscles inside the gluteus maximus and below the other gluteals serve two different purposes, again depending on position. During sitting they move the thigh out to the side, but if the leg is straight to the body, they rotate the upper leg, turning the knee outward.

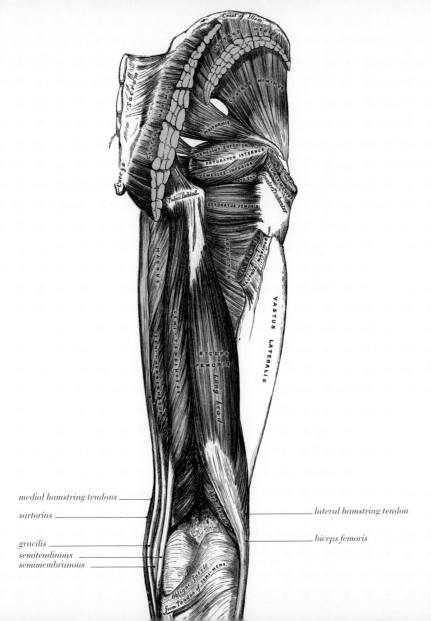

medial hamstring tendons

sartorius

gracilis

semitendinous

semimembranous

lateral hamstring tendon

biceps femoris

MUSCLES OF THE FRONT OF LOWER LEG

As with the arm. each joint in the leg is mainly controlled by the muscles above it. Since the movements of the ankle and foot are much more restricted by the nature of the joint than those of the wrist and hand. fewer muscles are required. None of the muscles at the front of the lower leg affects the movement of the knee. and they all run down the outer side of the leg. leaving the front and inner side of the tibia exposed. There are only four muscles and all of these are small. compared with the other muscles in the leg. The two extensor muscles straighten the toes (the extensor hallucis longus controlling the big toe. and the extensor digitorum longus pulling the others) and keep the foot perpendicular to the leg. The fibularis tertius and the tibialis anterior twist the sole of the foot sideways. raising it off the ground toward the outside and inside of the leg respectively. These two muscles, individually or together. also pull the top of the foot up toward the front of the leg. The upper surface of the foot has only a single active muscle (the extensor brevis digitorum) which helps the extensors in the leg to straighten the toes. This muscle is arranged diagonally across the foot to counteract the angle at which the long extensors pull.

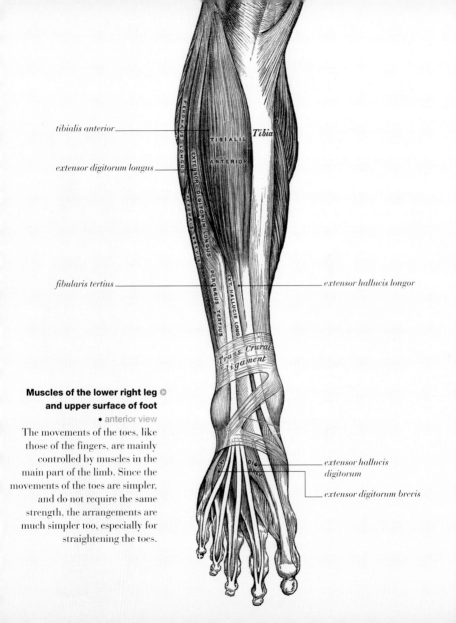

tibialis anterior

extensor digitorum longus

fibularis tertius

extensor hallucis longor

Muscles of the lower right leg ⊙
and upper surface of foot

● anterior view

The movements of the toes, like
those of the fingers, are mainly
controlled by muscles in the
main part of the limb. Since the
movements of the toes are simpler,
and do not require the same
strength, the arrangements are
much simpler too, especially for
straightening the toes.

extensor hallucis
digitorum

extensor digitorum brevis

MUSCLES OF THE BACK OF LOWER LEG

The muscles at the back of the calf are larger and more numerous than those of the front. The largest of them, the gastrocnemius, is one of only two major muscles of the lower leg to extend above the knee, passing between the hamstrings to connect to the condyles of the femur. It assists the muscles of the upper leg in bending the knee if the foot is fixed, but its main purpose is to raise the heel of the foot from the ground. It does this in association with the soleus, immediately beneath it, which connects to the upper parts of the tibia and fibula, the two muscles joining together at the bottom to form the Achilles tendon. The popliteus, which also connects to the base of the femur, is the first muscle to be activated when bending the knee, rotating the tibia slightly to ease the movement of the knee joint.

plantaris

achilles tendon

tendons of tibialis posterior

tendons of flexor digitorum longus

tendons of flexor hallucis longus

tendons of peronae longus and brevis

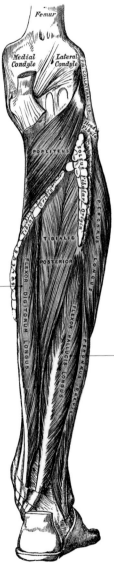

Femur

Medial Condyle

Lateral Condyle

POPLITEUS

TIBIALIS

POSTERIOR

flexor digitorum longus

flexor hallucis longus

◉ Deep muscles at the back of the right calf

The two flexor muscles, if contracted steadily, curl in first the toes and then, as far as possible, the sole of the foot before finally assisting the major muscles of the back of the calf in bringing the heel backward. The tibialis posterior, on the inside, and two fibularis muscles, on the outside, pull the sole of the foot up toward themselves. All three muscles also bring the heel up and backward.

◉ Surface muscles at the back of the right calf

The small plantaris muscle, which originates at the back of the knee joint, assists the gastrocnemius in pulling the heel up toward the back of the leg. Although this muscle contributes little in humans, it is the equivalent of a much larger and more powerful muscle found in the legs of most quadrupeds.

MUSCLES OF THE SOLE

The sole of the foot, like the palm of the hand, is covered by a strong fascia, thickest in the middle, which helps to protect the muscles and tendons from damage. The central part also has two vertical extensions, separating the three muscles in the shallowest layer from one another. Underneath the three layers shown is a fourth layer of muscles called the interossei. These work exactly like their namesakes in the hand.

Surface layer of muscle on the sole

The three muscles immediately beneath the plantar fascia are all connected to the calcaneus at the back of the foot. The flexor digitorum brevis, in the center, separates into four tendons; these interact with the tendons of the flexor digitorum longus in the calf before attaching to the tips of the four smaller toes. The adductor hallucis and adductor digiti minimi angle the toes, big and small respectively, to which they are attached out to the side of the foot.

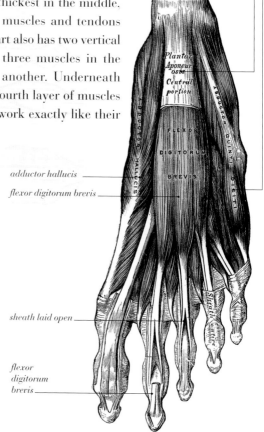

adductor minimi digiti

plantar fascia

adductor hallucis

flexor digitorum brevis

sheath laid open

flexor digitorum brevis

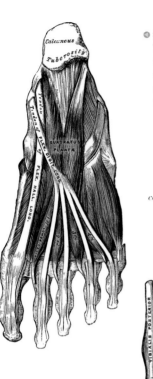

Second layer of muscle on the sole

The flexor digitorum longus gives rise to tendons that flex the three middle toes. The lumbricales—like their equivalents in the hand—also help with this action.

calcaneus

Third layer of muscle on the sole

The adductor hallucis transversus, connecting the metatarsals, pulls the toes together and deepens the arch of the foot. The adductor obliquus hallucis pulls the big toe toward the line of the foot.

adductor obliquus hallucis

adductor hallucis transversus

RED BLOOD CELLS

Red blood cells are vitally important: the hemoglobin they contain transports oxygen from the lungs to the rest of the body. This oxygen transportation is essential for the function and survival of the whole body. It is the hemoglobin that gives the cells their red color. Without the red cells, blood consists mainly of a yellowish liquid called plasma, which carries other nutrients needed by the body as well as a variety of other chemicals. Many of these are useful in different ways, but some are waste products of one kind or another, of which the most common are bicarbonate ions. These are produced when carbon dioxide (the waste product of oxygen metabolism) dissolves in the blood, and are carried to the lungs to be breathed out as carbon dioxide. The blood also contains white cells *(see p.134)*, and small bodies called platelets, which are responsible for forming blood clots at cuts to prevent blood loss.

The depth of the color of blood varies depending on the amount of oxygen it is carrying: the higher the level of oxygen, the brighter the red. However, veins can appear blue as a result of light diffraction by the skin; for the same reason, arteries seen through the skin also have a slight purple appearance. Red blood cells are normally disc shaped, with a large dent in each side, similar to two dinner plates back-to-back (see view B in the illustration), and should strictly be called corpuscles (not cells), as they have no nucleus. They are also extremely small (about $1/3,000$ of an inch/0.008 mm across), and huge numbers of them are required—around 2,500 billion per pint.

Red blood cells

• superior (A) and lateral (B) views

Red blood cells are often squashed or bent when passing through very small blood vessels under pressure, but the highly elastic membrane recovers its shape rapidly on returning to a larger vessel. One reason why the precise chemical make up of the blood must be controlled carefully by the body is that red blood cells are extremely sensitive to changes in their surroundings. If red blood cells are placed in water, they swell and become spherical, eventually disintegrating completely (C). The hemoglobin that gives them their red color will also pass out through the cell membrane, leaving them almost colorless and very difficult to see. If, on the other hand, they are put in a strong solution of sugar or salt (D), they become crinkled and spiky.

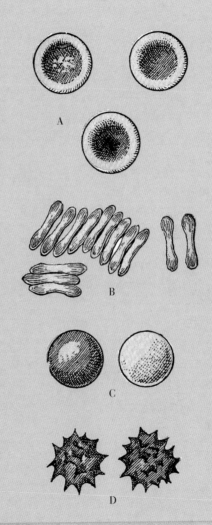

WHITE BLOOD CELLS

White blood cells are a vital part of the immune system and are far more varied and much rarer than the red corpuscles: a cubic millimeter (i.e. two millionths of a pint) of blood will typically contain four or five million red blood cells, but only between five and ten thousand white blood cells. Unlike red blood cells, white cells are also often found outside the bloodstream: individual white cells can be found almost anywhere in the body, combating infection and removing the remains of dead cells. White blood cells are produced in the bone marrow—along with red cells—and also in the lymphatic system. Lymphocytes are a group of white blood cell types, so called because they are found more in the lymphatic than the blood system, but they perform much the same range of functions as other white blood cell types. Other organs, most notably the thymus gland, which produces T cells *(see p. 286)*, are also involved in white cell production, adapting the cells produced elsewhere for specific purposes.

Some of the many different kinds of white blood cells

• highly magnified

In addition to their different appearance, the various types of white blood cell work to combat infections in different ways. Some are not directly involved in removing infections, but act instead to identify invaders and inform the killer cells of the infection's presence. Depending on the nature of the infection, different white cells can respond by eating the invader (phagocytosis), poisoning it using either histamines or oxygen, which is toxic to many bacteria, or by generating antibodies. Antibodies are specific to a particular infection, and can work in different ways. They act as a chemical message telling killer cells which cells are the infection, but antibodies for viral infections can also directly prevent the virus from infecting cells by making it larger and changing its shape.

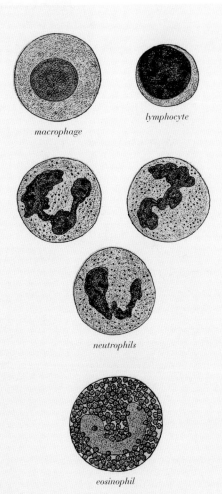

macrophage

lymphocyte

neutrophils

eosinophil

VEINS AND ARTERIES

In addition to transporting oxygen from the lungs to the rest of the body (especially the muscles) and removing waste carbon dioxide from the cells to be returned to and expelled from the lungs, the blood carries many other substances. In particular, the plasma of the blood supplies a cell with all the nutrients it requires, especially food in the form of glucose. Blood contains numerous other things: the white blood cells, which can be transported close to the site of an infection; small cells called platelets which, with other substances, form clots to prevent blood from flowing freely out of an open wound; waste materials from cells to be taken to the liver and kidneys for breakdown and disposal; and the many hormones used by different parts of the body to communicate with one another.

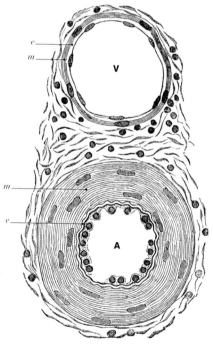

Cross-section through a typical ◉ vein (V) and small artery (A)
Both are lined with a thin membrane of endothelium (e). The muscular wall (m) of the artery is far thicker and stronger than that of the vein.

Contrary to expectation, the system of veins in the body as a whole is far larger than that of arteries in terms of the volume of blood it can hold, although the pulmonary circulation (to and from the lungs) has a roughly equal capacity in its local veins and arteries. Many veins, especially lower in the body but also many in the head, although none inside the brain, have simple valves to prevent blood from flowing in the wrong direction.

General pattern of circulation • blood ◉ flows clockwise around the diagram
If you consider the blood to start moving in the right side of the heart, it first passes through the lungs (the pulmonary circulation), where it disposes of carbon dioxide and collects oxygen. It then returns to the heart, and the stronger muscles of the left side pump it through the numerous arteries to the muscles and organs of the rest of the body (the systemic circulation), where the oxygen is gradually removed and replaced with waste carbon dioxide until it finally returns to the right-hand side of the heart to begin the process again.

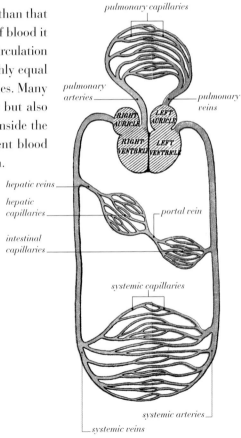

pulmonary capillaries

pulmonary arteries

pulmonary veins

RIGHT AURICLE   LEFT AURICLE

RIGHT VENTRICLE   LEFT VENTRICLE

hepatic veins

hepatic capillaries

portal vein

intestinal capillaries

systemic capillaries

systemic arteries

systemic veins

INSIDE THE HEART

The heart is an extremely complex arrangement of muscle designed to pump the blood to the rest of the body. Because of the nature of human circulation (from the heart to the lungs, back to the heart and out to the rest of the body), it is in fact two separate pumps working together, although the two pumps work in the same way. Blood fills the atria and first flows and is then forced into the ventricles by the contraction of the atria. The valve which separates the atria from its connecting ventricle then closes, and the ventricle contracts, pushing the blood out into the arteries.

Cross-section through the heart • showing three of the four chambers

The heart is made of a kind of muscle not found anywhere else in the body. It has two unique properties. First, while skeletal muscles require an electrical signal from the nerves to tell them to move, the muscle of the heart is (normally) self-activating. Second, it is active continuously from before birth until, usually, shortly before death. Other muscles can only be used for short periods before needing to rest.

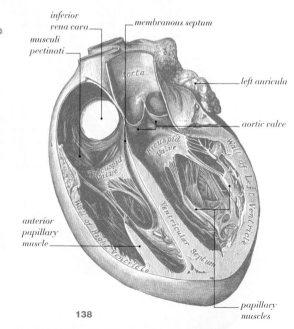

inferior vena cava

membranous septum

musculi pectinati

Aorta

left auricula

aortic valve

Bicuspid Valve

Tricuspid Valve

Wall of Left Ventricle

anterior papillary muscle

Ventricular Septum

Wall of Right Ventricle

papillary muscles

Right atrium and ventricle ○

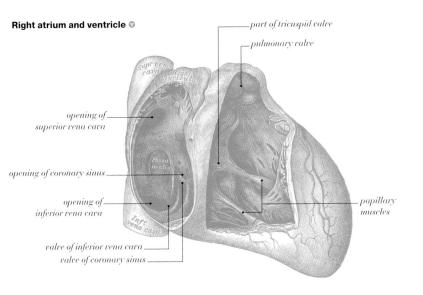

part of tricuspid valve
pulmonary valve
Supr. vena cava
Right auricula
opening of
superior vena cava
Fossa ovalis
opening of coronary sinus
opening of
inferior vena cava
Infr. vena cava
papillary
muscles
valve of inferior vena cava
valve of coronary sinus

Left atrium and ventricle ○

The muscles of the left ventricle, which push blood all the way around the body, are larger than those of the right, which only need to send blood to the lungs. The atrial walls are much the same on the right and left sides.

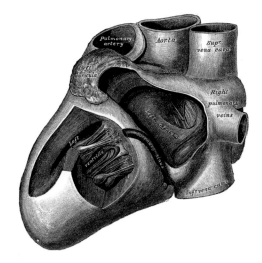

Pulmonary artery
Aorta
Supr. vena cava
Left auricula
Right pulmonary veins
Left ventricle
Left atrium
Infr. vena cava

THE HEART AND AORTA

Blood leaving the left ventricle of the heart enters the aorta, the largest and strongest blood vessel in the body. The first organ it supplies with blood is the heart itself, with the two coronary arteries branching from the main channel almost as soon as it exits the ventricle. The aorta then runs up and over the pulmonary arteries and top of the heart, with four major vessels (two carotid arteries to the head and two subclavian arteries, one to each arm) branching from it at the top of the aortic arch. The left carotid and subclavian arteries emerge directly from the aorta. Those for the right side of the body emerge joined together as a single large vessel (the brachiocephalic trunk), which carries the blood about an inch and a half across the top of the heart before branching into the right carotid and subclavian.

The heart and its major blood vessels ◯
The blue coloring of the pulmonary artery is correct, although the color is normally used for veins. It indicates low oxygen in the blood (which is never, in humans, actually blue). Because the lungs increase the level of oxygen in the blood, that in the pulmonary artery, leading away from the heart, has a very low oxygen content. Blood in the pulmonary veins (behind the heart in this view), which is returned from the lungs to the heart, contains the highest levels of oxygen found in the bloodstream anywhere in the body.

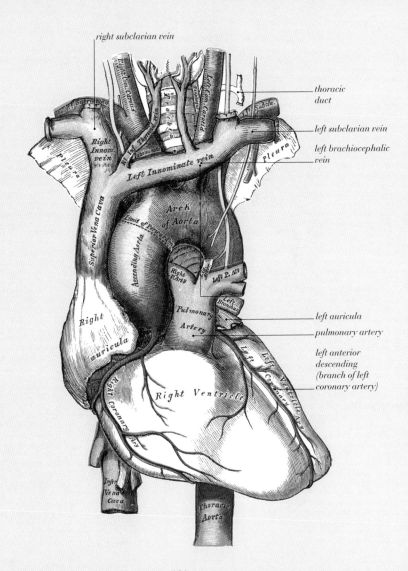

right subclavian vein

thoracic duct

left subclavian vein

left brachiocephalic vein

left auricula

pulmonary artery

left anterior descending (branch of left coronary artery)

BLOOD VESSELS AND THE HEART

The pulmonary arteries emerge from the heart and the pericardium as a single vessel, dividing in two underneath the arch of the aorta and branching as they take the blood to the lungs. The aorta, after completing its arch, runs down through the chest (where it becomes the thoracic aorta) and into the abdomen, (the abdominal aorta). While the blood leaves each ventricle of the heart in a single artery at the top, neither the pulmonary nor the main circulatory veins merge into a single vessel before entering the atria. Blood re-entering the heart from the main circulation comes in via the superior and inferior venae cavae at the top and bottom respectively of the right atrium, while the pulmonary veins enter the pericardium as four separate vessels leading to the left atrium.

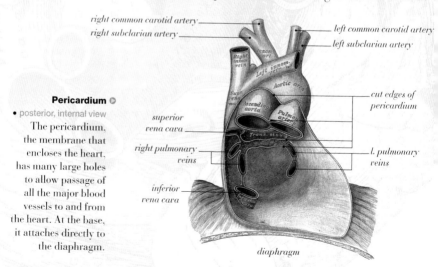

right common carotid artery

right subclavian artery

left common carotid artery

left subclavian artery

Pericardium ◉

• posterior, internal view

The pericardium, the membrane that encloses the heart, has many large holes to allow passage of all the major blood vessels to and from the heart. At the base, it attaches directly to the diaphragm.

cut edges of pericardium

superior vena cava

right pulmonary veins

l. pulmonary reins

inferior vena cava

diaphragm

Major blood vessels around the heart

• superior view

The pulmonary arteries emerge from the pericardium as a single vessel, splitting in two immediately underneath the arch of the aorta and branching rapidly as they take the blood to the lungs for replenishment of its oxygen.

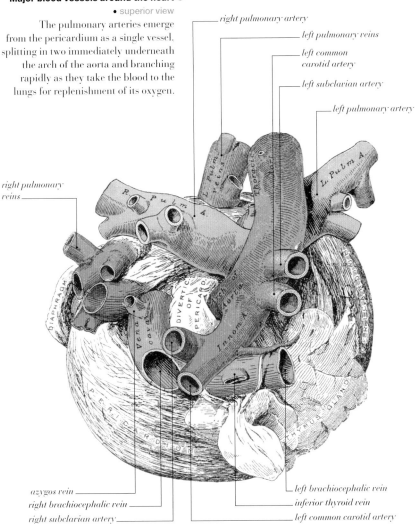

right pulmonary artery

left pulmonary veins

left common carotid artery

left subclavian artery

left pulmonary artery

right pulmonary veins

azygos rein

right brachiocephalic vein

right subclavian artery

left brachiocephalic vein

inferior thyroid rein

left common carotid artery

ARTERIES OF THE NECK

Common and external ◉
carotid arteries • skin
and some large muscles
removed, but showing
position of the blood
vessels relative to most
muscles of the neck
The internal jugular
vein, which has been
cut off in this view,
runs down the neck
lateral to the common
carotid artery.

external carotid artery

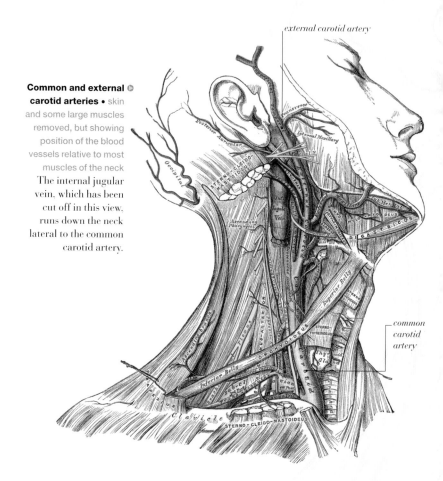

common
carotid
artery

The main vessels supplying oxygenated blood to the head are the two carotid arteries, which run up each side of the neck. Roughly level with the top of the thyroid cartilage *(see p. 254)*, the carotid splits into the external carotid artery, which supplies the jaw, nose, ears, and skin of the face and scalp; and the internal carotid artery, which supplies the brain and eyes.

Common and internal
carotid arteries
This shows the way in which the carotid artery runs vertically up the neck inside the head, while the throat and jaw extend forward past it. Also shown here is the right vertebral artery. This branches off the subclavian arteries, which supply blood to the upper limbs. The vertebral arteries are held in place up the sides of the vertebral column in holes in the transverse processes of the cervical vertebrae and supply blood to the vertebrae and to the brain. The two vertebral arteries join together inside the skull.

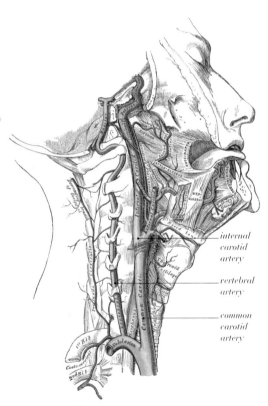

internal carotid artery

vertebral artery

common carotid artery

ARTERIES OF THE FACE AND SCALP

The face and scalp arteries are all branches of the external carotid artery *(see p. 144)*. The major branches are the lingual, facial, occipital, temporal, and maxillary arteries. Because of the relatively thin layer of muscle covering the bones of the head, all these arteries are extremely close to the surface, and many of them run between what little muscle there is and the skin. The temporal and occipital branches supply blood to the top and back of the scalp respectively, with the occipital artery supplying a number of muscles at the back of the neck. There are few muscles at the top of the skull so the branches are relatively small, supplying blood mainly to the skin and to the bone of the skull itself. The lingual and facial arteries between them supply blood to the muscles above and below the hyoid bone (including those of the tongue), to the salivary glands, to parts of the mouth and throat, and to much of the face.

Face and scalp arteries ○
The facial artery branches frequently, supplying all the muscles and other soft tissue outside the facial bones from the bottom of the jaw up to the eye sockets, which also receive some blood from a branch of the temporal artery. The muscles and organs inside the bones are, for the most part, supplied by either the lingual artery, which branches from the external carotid just below the facial, or by the maxillary artery *(see p. 148)*.

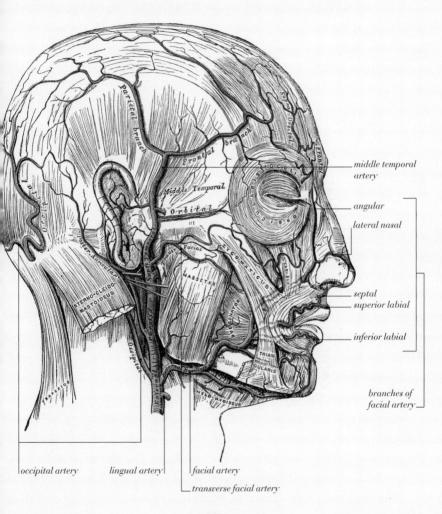

middle temporal artery

angular

lateral nasal

septal

superior labial

inferior labial

branches of facial artery

occipital artery

lingual artery

facial artery

transverse facial artery

ARTERIES INSIDE THE HEAD

The internal carotid arteries run closely parallel to their external pair, but slightly farther back and farther in, for some distance after the two have separated, before turning sideways and then up through an opening in the base of the skull. However, they do not branch at all over this distance, and supply blood only to the brain and, via the back of the eye socket, to the eyeball and some of the muscles within the socket.

Maxillary artery

This artery supplies those facial bones and muscles that are not fed by either the lingual or facial branches of the external carotid artery. In particular, it supplies blood to the nose and sinuses, most of the upper jaw, and parts of the pharynx and the cheeks, as well as the teeth of both the upper and lower jaws. The maxillary artery also provides the blood supply to some of the muscles inside the eye socket, and to the tear ducts, but not to the muscles controlling the eyelids (supplied by the facial artery) or to the eyeball itself.

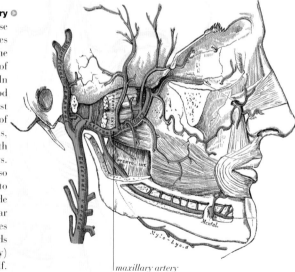

maxillary artery

Arrangement of arteries 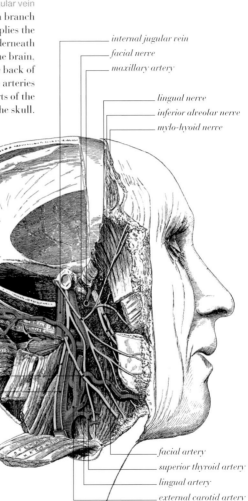 inside the lower part of the head

• with major nerves and internal jugular vein

Although the occipital artery is a branch of the external carotid and supplies the back of the scalp, it first passes underneath the part of the skull containing the brain, supplying blood to muscles at the back of the neck and base of the skull. The arteries here also supply blood to the parts of the cranial nerves that pass out of the skull.

internal jugular vein

facial nerve

maxillary artery

lingual nerve

inferior alveolar nerve

mylo-hyoid nerve

occipital artery

sub-occipital nerve

vertebral artery

hypoglossal nerve

internal carotid artery

facial artery

superior thyroid artery

lingual artery

external carotid artery

ARTERIES AND VEINS OF THE EYE

Blood to the eyeball and its immediate surroundings is supplied by the ophthalmic artery, a branch of the internal carotid artery. The first branch from the internal carotid (once it has entered the skull) is the tympanic artery, which supplies blood to the inner and middle parts of the ear. The internal carotid itself continues onward into the main cavity of the skull, entering very close to the optic nerve. Here, the internal carotid artery gives off a branch, the ophthalmic artery, which runs parallel to the optic nerve into the eye socket. The remainder of the internal carotid artery supplies blood to the brain.

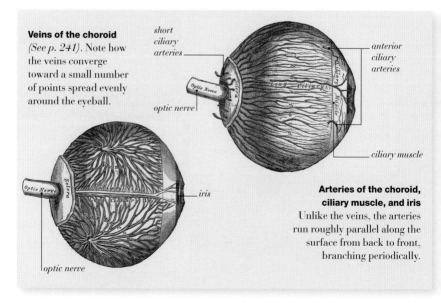

Veins of the choroid
(See p. 241). Note how the veins converge toward a small number of points spread evenly around the eyeball.

short ciliary arteries

optic nerve

anterior ciliary arteries

ciliary muscle

iris

optic nerve

Arteries of the choroid, ciliary muscle, and iris
Unlike the veins, the arteries run roughly parallel along the surface from back to front, branching periodically.

Arteries supplying the eyeball and its immediate surroundings

• superior view

The ophthalmic artery supplies the four recti muscles *(see p. 83)* for their entire lengths, but the two oblique muscles are provided with blood by the maxillary artery. As well as the muscles, the ophthalmic artery supplies the optic nerve, the lacrimal gland (the source of tears), and the eyeball itself. Various side branches also feed parts of the surrounding flesh and bone.

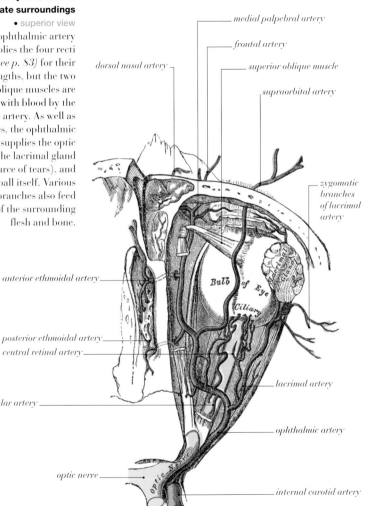

medial palpebral artery

frontal artery

dorsal nasal artery

superior oblique muscle

supraorbital artery

zygomatic branches of lacrimal artery

anterior ethmoidal artery

Bulb of Eye

Ciliary

posterior ethmoidal artery

central retinal artery

lacrimal artery

muscular artery

ophthalmic artery

optic nerve

Optic Nerve

internal carotid artery

ARTERIES OF THE BRAIN

The brain receives an enormous quantity of blood in relation to its mass and volume. In an adult, the brain weighs around 3 pounds (1.4 kg), slightly more in a man than a woman, and is only about 2% of the total body weight. However, the four arteries that feed it (the two internal carotids and the two vertebral arteries), deliver 15% of the blood pumped by the heart—one-seventh of the total amount. Unusually, the four arteries are all joined together, allowing blood to be transferred fairly freely from one section to another through the communicating arteries.

Middle cerebral artery ○

This artery is the larger of the two major branches of the internal carotid artery inside the brain. Via its own branches it supplies blood to nearly half of each side of the brain.

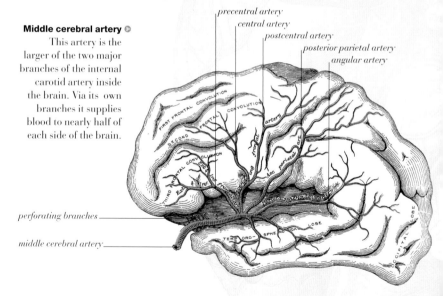

precentral artery
central artery
postcentral artery
posterior parietal artery
angular artery

perforating branches

middle cerebral artery

Major arteries supplying blood to the brain • view from below

Part of the cerebellum has been removed to show the occipital lobe above. The two vertebral arteries entering the skull near the back send branches to the cerebellum and midbrain and then merge into the basilar artery, the single artery that runs across the pons. This has numerous small side branches, which feed the pons and cerebellum, before dividing into the two posterior cerebral arteries.

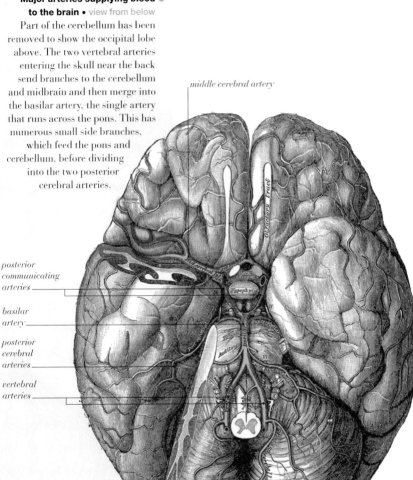

middle cerebral artery

posterior communicating arteries

basilar artery

posterior cerebral arteries

vertebral arteries

ARTERIES OF THE RIB CAGE

The ribs and sternum and the muscles around them are supplied with blood by the internal thoracic arteries, which are branches of the first part of the subclavian artery in the neck. The internal thoracic arteries run vertically down the chest behind the costal cartilages *(see p. 33)* of the rib cage, an inch or two (3–5 cm) out from the side of the sternum. Roughly level with the base of the sternum, each internal thoracic splits into two branches of about equal size. The internal thoracic then becomes the musculophrenic artery, which supplies blood to the lower ribs, and the superior epigastric artery, which runs within the sheath of the rectus abdominis muscle *(see p. 102)*, providing much of the supply of blood for this and the other superficial muscles of the abdomen. Smaller branches farther up the internal thoracic provide some of the blood supply to the pectorals, the diaphragm, and the pericardium around the heart, as well as to the mammary glands in the breasts *(see p. 304)*.

Arteries supplying the ribs and abdominal muscles ⊙

• outer layers of muscle of the abdomen removed to show epigastric arteries

The superior and inferior epigastric arteries supply the abdominal muscles from opposite ends. The inferior epigastric rises up from the external iliac artery inside the pelvis. The brachiocephalic trunk, a short section of artery that branches off the aorta and almost immediately splits into the carotid and subclavian arteries, is unusual in that it (normally) only exists on the right-hand side.

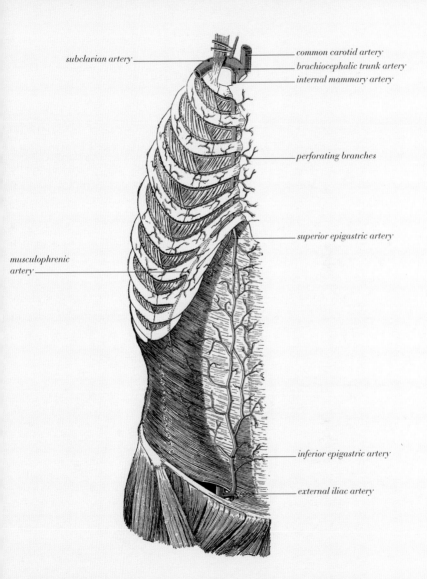

subclavian artery

common carotid artery

brachiocephalic trunk artery

internal mammary artery

perforating branches

superior epigastric artery

musculophrenic artery

inferior epigastric artery

external iliac artery

ARTERIES OF THE SHOULDER

The shoulders and the muscles around them are supplied by three separate arteries on each side of the body. At the back, the subscapular (or transverse scapula) and transverse cervical arteries branch off the front of the subclavian artery, a little distance after the internal thoracic. The two follow similar courses outward and backward—the subscapular artery being the farther back of the two—and supply blood to all the muscles around the back of the shoulder and upper arm, including much of the large deltoid and trapezius muscles. At the front, the muscles around the shoulder and upper arm receive blood from the axillary artery and its branches.

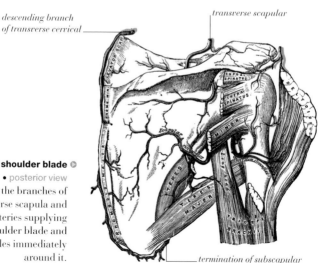

descending branch of transverse cervical

transverse scapular

Right shoulder blade ○

• posterior view

This shows the branches of the transverse scapula and cervical arteries supplying the shoulder blade and muscles immediately around it.

termination of subscapular

Axillary artery and its branches

The pectorals (including part of the deltoid) connecting the upper arm to the front of the chest are supplied by the axillary artery. This is the continuation of the subclavian artery once it has passed into the axilla (armpit), and it extends all the way to the hand. Below the axilla the artery is known as the brachial.

axillary artery

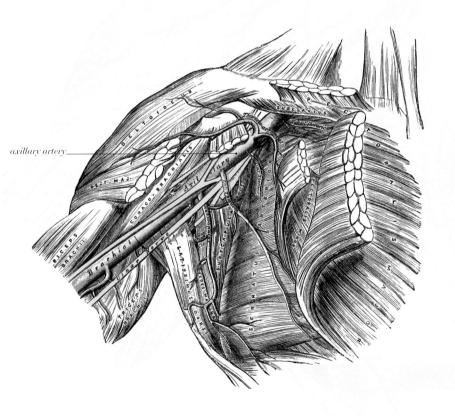

ARTERIES OF THE UPPER ARM

The muscles of the upper arm are supplied with blood by the brachial artery, the continuation of the axillary artery. Along with the nerves of the arm and the basilic vein, it runs along the inside of the biceps muscle from the shoulder to the elbow. The artery is close to the surface for its entire length. For some distance below the shoulder it is close to the surface on the side of the arm that rests against the rib cage. Even so, it is rarely damaged because it is protected by the main part of the arm on one side and the entire body on the other. Lower down, however, it moves around toward the front of the arm. While it is no closer to the surface here than farther up, it is much more exposed, and thus easier to damage.

Owing to its size and closeness to the surface, the brachial artery is well-suited to use for medical purposes. The artery is also used to measure blood pressure because the reading can be made without being affected by muscle (or much else) between the artery and the cuff of the blood pressure meter.

Brachial artery ⊙

The muscles of the upper arm all receive their blood from either the brachial artery or long branches of the transverse arteries *(see p. 156)* that supply the back of the shoulder. There is no major vessel equivalent to the brachial artery down the back of the arm.

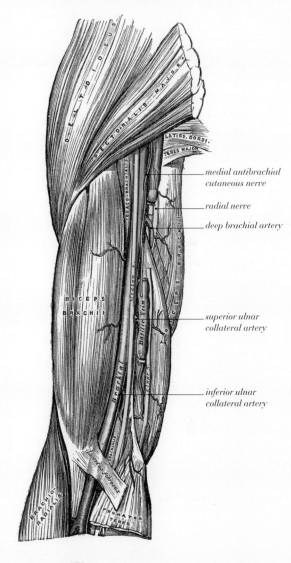

medial antibrachial cutaneous nerve

radial nerve

deep brachial artery

superior ulnar collateral artery

inferior ulnar collateral artery

ARTERIES OF THE FOREARM AND HAND 1

Just below the elbow the brachial artery divides into two roughly equal branches. They run along the two bones of the forearm and are therefore known as the radial and ulnar arteries. The latter (which is the slightly larger of the two) passes underneath the pronator and flexor muscles to run down the inside edge of the forearm. The radial artery.

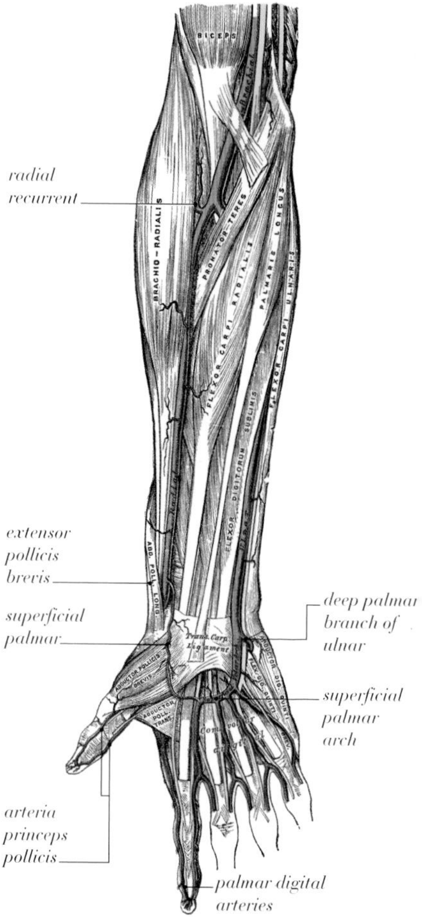

Arteries of the right forearm and hand • in relation to the muscles, nerves, and tendons The palm and fingers are supplied with blood by two different sets of vessels. Shown here is the upper set branching off the superficial palmar arch, which forms the extreme end of the ulnar artery.

radial recurrent

extensor pollicis brevis

superficial palmar

arteria princeps pollicis

deep palmar branch of ulnar

superficial palmar arch

palmar digital arteries

which remains close to the surface down the upper part of the forearm, is often visible toward the outside edge of the arm. At the wrist, however, the radial artery, from which the pulse is normally taken, travels underneath the strong fascia and muscles and disappears into the pad of flesh at the base of the thumb.

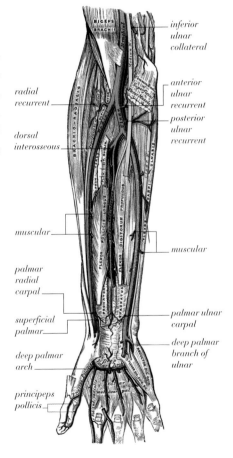

Arteries of the right forearm and hand • deep view

The lower set of blood vessels in the hand are fed by the deep palmar arch, a loop connecting the radial artery to a smaller branch of the ulnar. From this, arteries are also sent through the palm of the hand, supplying blood to the bones and joining the arteries on the back surface.

inferior ulnar collateral

anterior ulnar recurrent

posterior ulnar recurrent

radial recurrent

dorsal interosseous

muscular

muscular

palmar radial carpal

superficial palmar

palmar ulnar carpal

deep palmar branch of ulnar

deep palmar arch

principeps pollicis

ARTERIES OF THE FOREARM AND HAND 2

Since there is relatively little muscle on the back of the forearm and hand, these areas do not require a major blood vessel; instead, they are supplied by various smaller branches. Some of these descend from the arteries in the back of the shoulder and the muscles at the back of the upper arm, while others—particularly in the hand—come around the bones or between them from the radial and ulnar arteries. The most notable of these are the posterior carpal arteries, which run around the sides of the wrist from the radial and ulnar arteries at the front and join together to supply the arteries that run along the length of the back of the hand. Three of these, those for the three long fingers, are then joined through the spaces between the metacarpal bones by the perforating arteries, transferring blood from the deep palmar arch on the other side of the hand *(see p. 161)*.

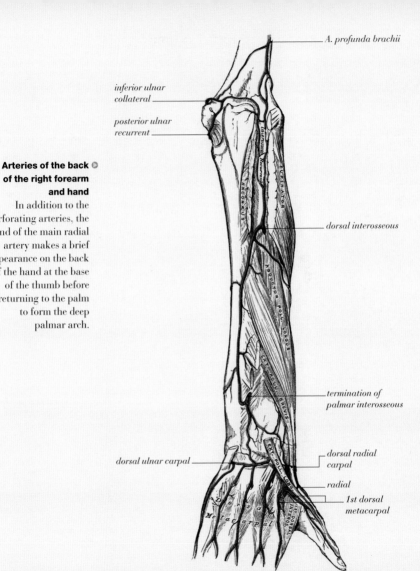

Arteries of the back of the right forearm and hand

In addition to the perforating arteries, the end of the main radial artery makes a brief appearance on the back of the hand at the base of the thumb before returning to the palm to form the deep palmar arch.

A. profunda brachii

inferior ulnar collateral

posterior ulnar recurrent

dorsal interosseous

termination of palmar interosseous

dorsal ulnar carpal

dorsal radial carpal

radial

1st dorsal metacarpal

THE ABDOMINAL AORTA

When it passes through the diaphragm, the thoracic aorta becomes the abdominal aorta. The first lateral branches of the abdominal aorta are small vessels that supply the inferior surface of the diaphragm and the adrenal glands. The inferior phrenic arteries supply blood to the diaphragm, and can branch either directly from the abdominal aorta or from one of its branches. However, within an inch of the inferior surface of the diaphragm, a major midline branch, the celiac trunk, arises to supply blood to the stomach, spleen, liver, and part of the pancreas. A little distance inferior to the celiac trunk another midline branch arises, the superior mesenteric artery *(see p. 168)*, which supplies blood to the midgut. At the level of the superior mesenteric artery two lateral branches, the renal arteries, arise from the aorta to supply the kidneys. The next inferior midline branch is known as the inferior mesenteric artery, which supplies the hindgut region.

Abdominal aorta and its ⊙
major branches • digestive
system removed
Slightly above the pelvis, the abdominal aorta splits into two equal-sized vessels: the common iliac arteries. A little lower down, each of these divides into the internal iliac artery, which supplies the muscles and organs inside the pelvic cavity, and the larger external iliac artery, which subsequently runs down the leg.

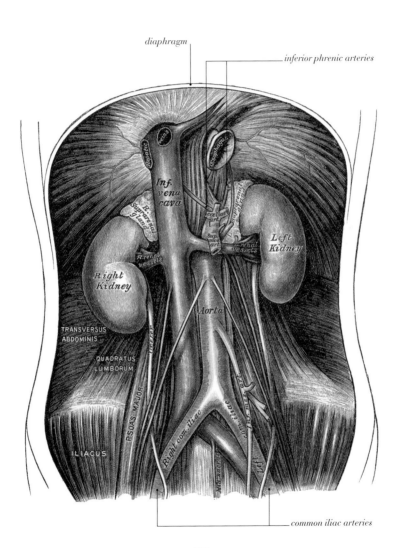

diaphragm

inferior phrenic arteries

common iliac arteries

Inf.
vena
cava

R.
Suprarenal
Gland

Right
Kidney

R. renal
vessels

TRANSVERSUS
ABDOMINIS

QUADRATUS
LUMBORUM

PSOAS MAJOR

ILIACUS

Ureter

Aorta

Right com. iliac

Coeliac art.

*Sup.
mes.
art.*

*L. renal
vessels*

*Left
Kidney*

Suprarenal Gland

165

ARTERIES OF THE UPPER ABDOMEN

The first three major arteries to come off the abdominal aorta do so as a group, via a short (¹/₂ inch/10 mm) vessel called the celiac axis which separates from the aorta before branching into three. These arteries are the gastric (to the stomach), hepatic (to the liver), and splenic (supplying the spleen and pancreas). In a fetus or a young child, the largest of these is the hepatic, but well before puberty this is overtaken in size by the splenic artery. Two of these three arteries are not quite as dedicated as their names would suggest. The hepatic artery soon splits, with one branch supplying the liver while others take blood to the pancreas, duodenum, and outer curve of the stomach, as well as the greater omentum of the peritoneum. The splenic artery supplies much of the pancreas before dividing into two major branches: one continues to the spleen while the other curves around the outside of the stomach to join up with the branch of the hepatic artery.

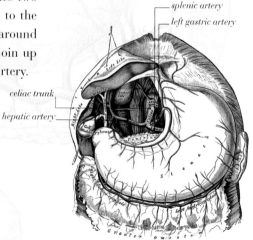

splenic artery

left gastric artery

celiac trunk

hepatic artery

Arteries of the upper ◉
abdomen • liver removed
but other organs in place
The greater omentum, at
the base, is a fold of the
peritoneum, the large two-
layered membrane that protects
the contents of the abdomen
and allows the organs to move
relative to one another.

Arteries of the upper abdomen ⊙

- peritoneum removed and
  stomach raised up

This view shows the arrangement
of blood vessels to the organs
behind, including the great loop
formed by the gastric branches of
the hepatic and splenic arteries.

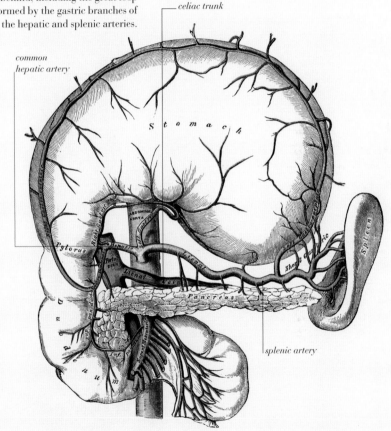

celiac trunk

*common
hepatic artery*

splenic artery

ARTERIES OF THE INTESTINES

The largest branches of the abdominal aorta are the celiac trunk, the two renal arteries *(see p. 164)*, and the superior mesenteric. As the superior mesenteric artery runs anterior between the pancreas and the duodenum of the small intestine, it is already producing many branches and diminishing rapidly in diameter. The first of these branches supplies blood to the duodenum and to the curved head of the pancreas; the other branches on this side (of which there are around fifteen) run roughly parallel to each other, and feed the small intestine. These parallel arteries repeatedly divide, with many of these branches combining with a branch from one of the neighboring arteries to

Superior mesenteric ○ artery • upper digestive organs removed and intestines rearranged to display the network of branches

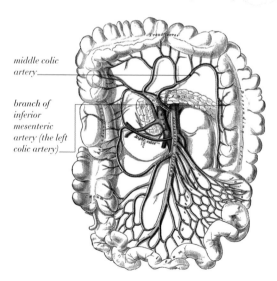

middle colic artery

branch of inferior mesenteric artery (the left colic artery) form a series of arch shapes, becoming steadily smaller and more numerous as they approach the small intestine. The rest of the large intestine is supplied by the inferior mesenteric artery, which branches directly from the aorta a few inches below the superior.

On the other side of the superior mesenteric artery are two major branches with a smaller one between. The upper two of these branches supply blood to the ascending and transverse parts of the colon; the lowest one feeds the cecum, appendix, and last part of the ileum. They all branch in similar fashion to the arteries feeding the small intestine, but less spectacularly.

Inferior mesenteric artery ◐
• some branches of the superior mesenteric artery visible

The hemorrhoidal arteries provide the supply of blood to the muscles of the anus, although it is the veins, not the arteries, that become a problem in the painful medical condition known as hemorrhoids or piles.

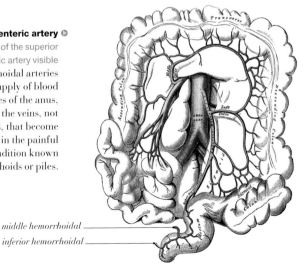

middle hemorrhoidal _____

inferior hemorrhoidal _____

ARTERIES OF THE PELVIS

The interior of the pelvis is supplied with blood by the branches of the two internal iliac arteries. Having separated from the common iliac, the internal iliac itself splits into two major branches only a couple of inches farther down. Of these, the posterior trunk supplies some of the muscles in the hip and back of the thigh, and also joins the lower part of the blood supply of the vertebral column, just above the sacrum. The anterior trunk itself runs almost straight downward, giving off numerous branches forward and to the side: these carry blood to the bladder and ureters *(see p. 280),* and to the internal sexual organs, as well as providing part of the supply of blood to the rectum and anus. Just before passing the muscular layer at the base of the pelvis, the anterior trunk splits into two further branches. The larger of these is the inferior gluteal artery, which contributes to the blood supply in the hip and thigh. The other is the internal pudendal artery, which supplies blood to the muscles of the pelvic floor, the perineum, and external genitalia.

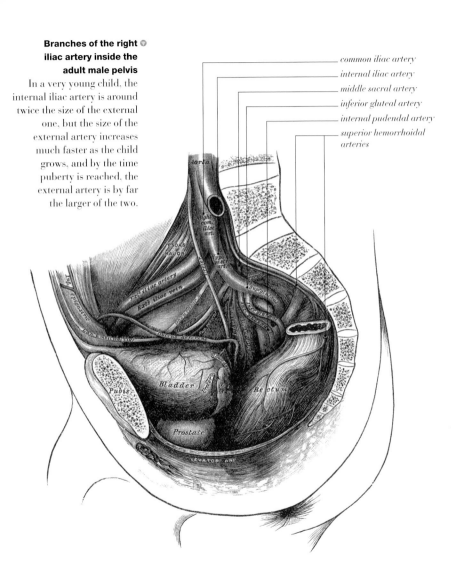

Branches of the right iliac artery inside the adult male pelvis

In a very young child, the internal iliac artery is around twice the size of the external one, but the size of the external artery increases much faster as the child grows, and by the time puberty is reached, the external artery is by far the larger of the two.

common iliac artery

internal iliac artery

middle sacral artery

inferior gluteal artery

internal pudendal artery

superior hemorrhoidal arteries

Aorta

Right com. iliac art.

PSOAS MAJOR

Hypogastric art.

Spermatic artery

Ext. iliac vein

Vas deferens

Inf. Epigastric

Obliterated hypogastric

Bladder

Rectum

Pubis

Prostate

LEVATOR ANI

ARTERIES OF THE HIP, BUTTOCKS, AND BACK OF THIGH

The external iliac artery runs anteriorly and to the side from the point where the common iliac splits in two. It has only two branches of any size before it passes into the leg, where it becomes the femoral artery. The first is the inferior epigastric artery *(see p. 155)*, which supplies part of the abdominal wall. The second is the deep circumflex iliac artery, which runs upward and outward to the crest of ilium (the highest part of the pelvis; see p. 45), where it provides some branches to the lower end of the abdominal muscles before passing down and back to join the branches of the internal iliac artery in supplying the muscles of the back of the hip.

Most of the supply of blood to the muscles of the buttocks and hips comes from the lower branches of the internal iliac artery and the four major branches into which the posterior trunk is divided. While the upper three of these deliver primarily to the inner muscles and the sacrum and coccyx, the gluteal artery, the lowest and largest branch, provides the blood supply not only to the gluteal muscles, but also to the components of the hip joint itself.

Arrangement of the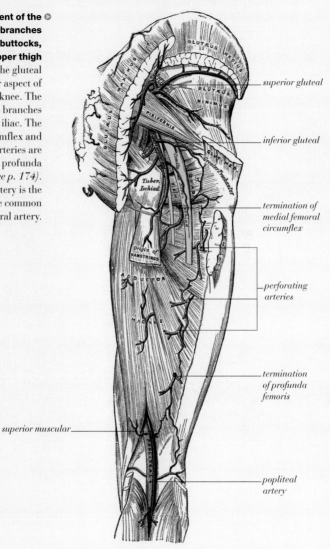
many arterial branches
supplying the buttocks,
hip, and upper thigh

Blood supply to the gluteal
region and posterior aspect of
the thigh and knee. The
gluteal arteries are branches
of the internal iliac. The
medial circumflex and
perforating arteries are
branches of the profunda
femoris artery *(see p. 174)*.
The popliteal artery is the
continuation of the common
femoral artery.

superior gluteal

inferior gluteal

*termination of
medial femoral
circumflex*

*perforating
arteries*

*termination
of profunda
femoris*

superior muscular

*popliteal
artery*

THE FEMORAL ARTERY

The external iliac artery becomes the femoral artery where it passes under the inguinal ligament, and down into the thigh. It runs roughly parallel with the femur, coming out of the pelvis almost exactly level with the hip joint. Here it is protected by a strong sheath of aponeurotic fibers. It is, however, unprotected by any muscles: only this sheath and the normal layers of fascia *(see p. 122)* lie between it and the skin, although the superficial fascia of fatty tissue is relatively thick on the thighs. About two-thirds of the way down the thigh, the artery passes through an aperture in the adductor magnus muscle *(see adductor muscles p. 123)*, and re-emerges as the popliteal artery at the back of the knee *(see p. 176)*.

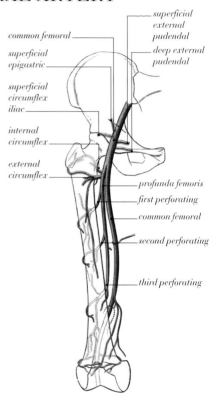

common femoral
superficial epigastric
superficial circumflex iliac
internal circumflex
external circumflex

superficial external pudendal
deep external pudendal

profunda femoris
first perforating
common femoral
second perforating
third perforating

◉ **Femoral artery and main and perforating branches**

● anterior view, muscles removed

Femoral artery and its surroundings

The first branch of the femoral artery passes straight back up over the front of the inguinal ligament, supplying blood to the skin and external oblique muscle up to around the level of the navel. By far the largest branch is the profunda femoris. It passes back between the muscles of the upper thigh, gives off "perforating" arteries (so called because they pass through the adductors), and provides the blood supply for the muscles at the back of the knee.

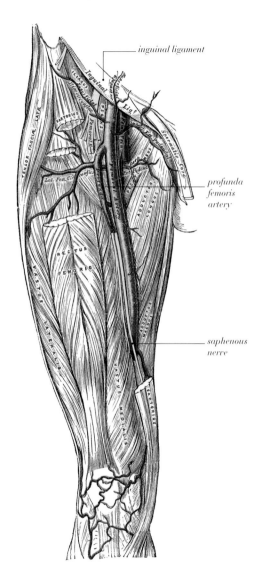

inguinal ligament

profunda femoris artery

saphenous nerve

ARTERIES OF THE LOWER LEG

Emerging from the back of the knee, the femoral artery becomes the popliteal artery. This has a number of small branches supplying blood to the muscles at the back of the knee, including the lower end of the muscles of the back of the thigh. It splits just below the knee into the posterior and anterior tibial arteries. The anterior tibial artery immediately passes forward through the gap between the bones of the lower leg, and runs down the front of the leg to the foot. The posterior tibial artery, which is much larger, runs from this junction down the flexor muscles in the back of the calf, producing one major branch near the top, the peroneal artery, which runs deep in the leg close to the fibula down to the ankle.

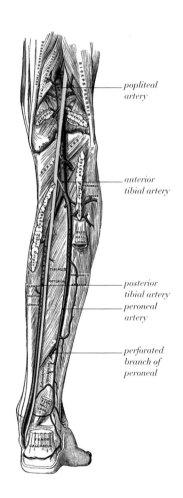

popliteal artery

anterior tibial artery

posterior tibial artery

peroneal artery

perforated branch of peroneal

Anterior tibial artery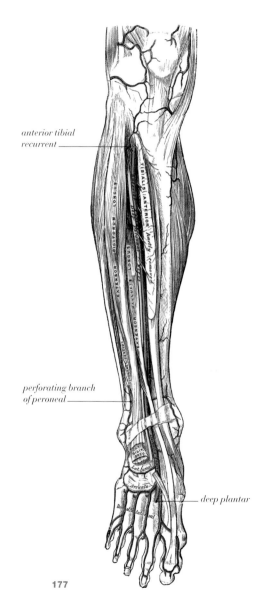

• some of the muscles at the front of the leg (which this artery supplies) removed

The anterior tibial artery lies beneath the muscles to protect it from damage. Its main function is to supply blood to the front of the leg and upper surface of the foot (where it becomes the dorsalis pedis), but some early branches head upward to the knee. The two genicular arteries are branches from the popliteal, passing around the sides of the knee to supply the front of the knee.

Popliteal, posterior tibial, and peroneal arteries

These supply the muscles in the back of the leg from the knee down. The posterior tibial artery is still a major blood vessel when it reaches the foot, but the peroneal artery, the last branches of which supply some parts of the heel bone and Achilles tendon, is much smaller.

anterior tibial recurrent

perforating branch of peroneal

deep plantar

177

ARTERIES OF THE FOOT

The sole of the foot receives its main supply from the posterior tibial artery, although parts of the heel are supplied by the peroneal artery. On reaching the sole of the foot, the posterior tibial artery divides in two. The medial plantar artery runs along the arch of the foot to the outside edge of the big toe, while the much larger lateral plantar artery runs diagonally across to the base of the little toe before sweeping back across the ball of the foot sending branches to the toes. The upper surface of the foot is supplied with blood by the dorsalis pedis, the continuation of the anterior tibial artery past the ankle (shown opposite). This sends two branches—the tarsal and metatarsal arteries—sideways across the foot and one—the communicating artery—down between the metatarsal bones to the sole of the foot. The big toe is served by the last section of the dorsalis pedis, while the other toes receive parallel branches from the metatarsal artery.

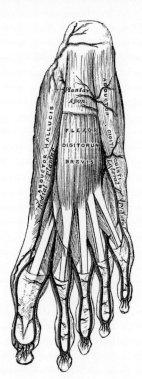

◁ Shallow arteries on the sole of the foot

This shows the supply of blood from the peroneal artery to the heel as well as the branches of the plantar arteries that feed the surface muscles and skin.

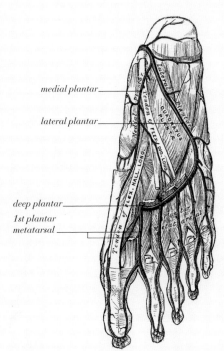

medial plantar

lateral plantar

deep plantar

1st plantar metatarsal

Deeper arteries in the ◉ sole of the foot

• muscles removed to show the two main plantar arteries

The communicating branch of the dorsalis pedis forms a loop with the extreme end of the lateral plantar artery, supplying blood to the underside of the toes.

SPINAL ARTERIES AND VEINS

The vertebrae receive blood from one of two different routes, depending on their position in the spine. The cervical vertebrae are supplied by the vertebral arteries that run up the side of the neck *(see p. 144)*, while the lower vertebrae are served by branches from the aorta and, within the pelvis, by the iliac arteries. These branches enter the vertebral canal through the gaps between vertebrae and supply blood to the spinal cord, vertebrae, and surrounding tissue. Unusually, there are no valves in the system of spinal veins.

Horizontal cross-section ⊙
through a thoracic vertebra
The vertebrae, like the bones of the skull, have large veins passing through the soft spongy tissue in the center of the bones. The spinal veins are part of a complex network (shown in blue) connecting veins inside the vertebral column with those outside it, both around the main body of the vertebrae and running up both sides of the spinous processes.

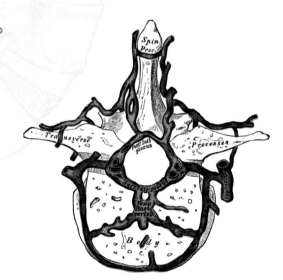

Vertical cross-section through the spine in the thoracic region

This shows the complex connections between the plexuses (or networks) of veins in different parts of the vertebral column.

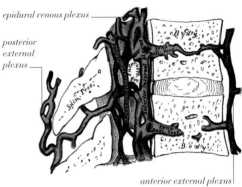

epidural venous plexus

posterior external plexus

anterior external plexus

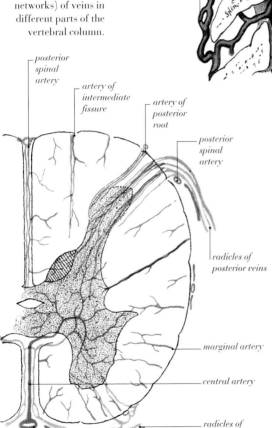

posterior spinal artery

artery of intermediate fissure

artery of posterior root

posterior spinal artery

radicles of posterior veins

marginal artery

central artery

radicles of anterior veins

anterior spinal artery

Cross-section of the spinal cord

• showing arrangement of veins and arteries

Veins and arteries that actually pass into the spinal cord tend to head fairly directly inward to the center until the gray matter is reached.

VEINS OF THE HEAD AND NECK

Since the head receives such a high proportion of the blood, it necessarily requires a large network of veins to remove it. There are, in fact, two largely separate networks, although there is some connection between them. The veins from the scalp and the deeper parts of the face—the top of the throat, the back of the jaw, and the nasal cavity—combine to form the external jugular vein. The internal jugular vein receives blood from the brain and from areas of skin and muscle at the front of the face. There are two additional jugular veins on either side of the neck. At the back is the posterior jugular, which collects blood from the surface muscles at the upper part of the back of the neck before joining the external jugular. Running close to the front is the anterior jugular. This starts at the junction of several veins beneath the tongue (near the hyoid bone) and runs all the way to the base of the neck before joining either the external jugular or the subclavian vein.

Arrangement of major veins on
the outside of the head and neck
Most of the blood from the head
drains into the main internal and
external jugular veins, while the
posterior and anterior veins collect
blood from the extreme front and
rear parts of the head.

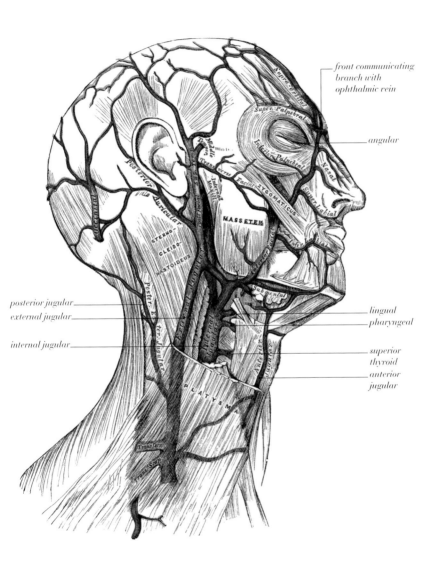

front communicating branch with ophthalmic vein

angular

posterior jugular

external jugular

internal jugular

lingual

pharyngeal

superior thyroid

anterior jugular

VEINS OF THE SKULL

In addition to their relatively large size, the veins in and immediately around the brain differ from those elsewhere in the body by their complete lack of any muscular covering. To cope with the very large quantities of blood used by the brain, there are two sets of veins around the brain that are not found elsewhere in the body: the venous sinuses of the dura mater (the membrane protecting the brain) and the veins of the diploe (the spongy bone found in the skull).

Veins of the diploe • hard ◉ outer layer of bone removed The bones of the skull, like most others, consist of two hard layers of compact bone with a layer of softer spongy material, (the diploe) in between. However, the skull—and vertebrae, where a similar arrangement exists—is unique in having a number of large channels through this softer layer in which veins run, carrying blood from the brain down to the jugular.

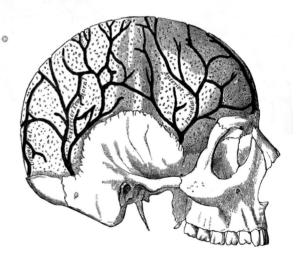

Horizontal cross-section of the skull

• showing venous sinuses in the skull

The transverse venous sinuses in particular hold a very large quantity of blood which is drained into the internal jugular.

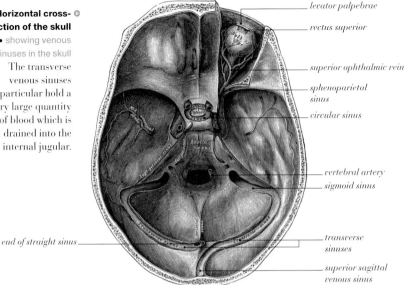

levator palpebrae

rectus superior

superior ophthalmic vein

sphenoparietal sinus

circular sinus

vertebral artery

sigmoid sinus

transverse sinuses

superior sagittal venous sinus

end of straight sinus

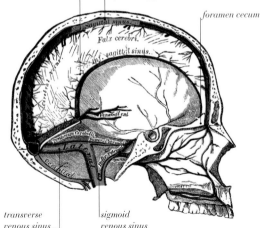

inferior sagittal venous sinus

superior saggital venous sinus

foramen cecum

transverse venous sinus

sigmoid venous sinus

Vertical cross-section of the skull

• showing sinuses of the dura mater

The veins inside the skull itself drain into a series of large cavities called sinuses within and around the brain. The sinuses, like the veins within the brain, have a thin lining of endothelial cells but no other coating.

SUPERFICIAL VEINS OF THE ARM

The wrist and hand receive relatively large quantities of blood for their size—particularly for the amount of muscle tissue they contain—although not nearly as much as the brain. The veins that provide the drainage to remove this blood are of two types: superficial and deep. The superficial veins lie immediately beneath the skin, as do the superficial fascia muscles—many of which can easily be seen. The deep veins of the arm and hand remain beneath the surface, running parallel to several of the major arteries, most notably the deep palmar arch, and the radial, ulnar, brachial *(see p. 160)*, and axillary arteries. The axillary vein is, in fact, the upward continuation of the brachial vein.

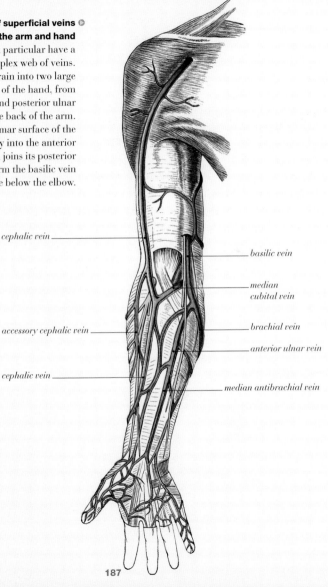

Arrangement of superficial veins
draining blood from the arm and hand
The fingers in particular have a
surprisingly complex web of veins.
These in most cases drain into two large
junctions on the back of the hand, from
which the radial and posterior ulnar
veins run up the back of the arm.
The veins on the palmar surface of the
hand drain mainly into the anterior
ulnar vein, which joins its posterior
companion to form the basilic vein
a short distance below the elbow.

cephalic rein

basilic vein

*median
cubital vein*

accessory cephalic vein

brachial vein

anterior ulnar vein

cephalic rein

median antibrachial rein

MAIN VEINS OF THE TORSO

The arrangement of veins in the torso, or main body, is very similar to the arrangement of the primary arteries, except that the veins are invariably larger than their equivalents. Since veins are under lower pressure than arteries (and are naturally more "stretchy" because they have less muscle to squeeze them), they can be used as a form of blood reservoir.

The veins of each arm—along with those of the shoulder and lower neck—drain into the subclavian vein of the same side. These combine with the jugular veins to form the left and right brachiocephalic veins, the final junction of which becomes the superior vena cava. Similarly, the femoral veins of the legs drain into the external iliac veins, which join the internal iliac veins to form two common iliac veins. In turn these join, along with the middle sacral vein (which drains blood from the sacrum and coccyx), to form the inferior vena cava into which the veins from the kidneys and liver also drain.

At the back of the chest, the azygos veins, which collect the blood from the intercostal muscles, are connected to both the inferior and superior venae cavae, allowing obstructions to be bypassed by the blood. This means, for instance, that should one of the venae cavae become blocked, some blood can flow through the azygos veins to reach the heart.

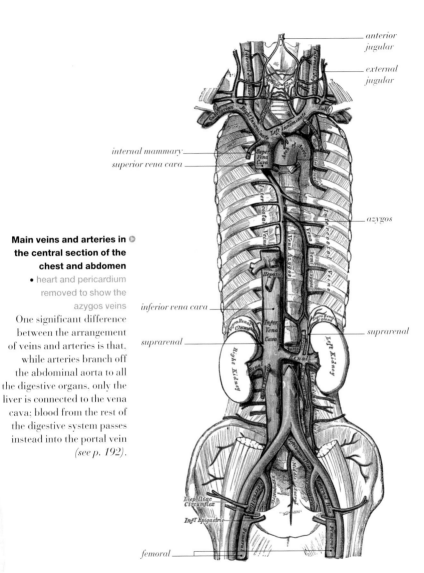

Main veins and arteries in
the central section of the
chest and abdomen

• heart and pericardium
removed to show the
azygos veins

One significant difference
between the arrangement
of veins and arteries is that,
while arteries branch off
the abdominal aorta to all
the digestive organs, only the
liver is connected to the vena
cava: blood from the rest of
the digestive system passes
instead into the portal vein
(see p. 192).

anterior jugular

external jugular

internal mammary

superior vena cara

azygos

inferior vena cara

suprarenal

suprarenal

femoral

VEINS OF THE LEG

The veins of the leg, like the veins of the arm, can be divided into the superficial (shallow) veins, shown here, and the deep veins, which remain close to the equivalent arteries *(see p. 172–177)*. However, the leg veins, in addition to being larger than those in the arms, have a thicker and stronger muscular wall; they also have many more valves than in equivalent veins elsewhere in the body. Both of these differences are necessary to counteract the much greater pressures that arise in the veins in the leg—especially near the foot—since they are in effect part of a column of blood stretching from the foot all the way up to the right atrium of the heart. Varicose veins, which are caused by stretching of the vein walls as a result of the failure of valves, are most often a problem in the legs for this reason.

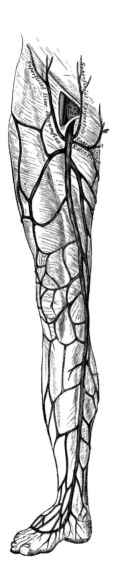

Superficial veins of the leg
• lateral view

The larger of the two main superficial veins is the great saphenous vein. which can have up to twenty valves to help reduce the pressure between the foot and the top of the thigh. where it drains into the femoral vein.

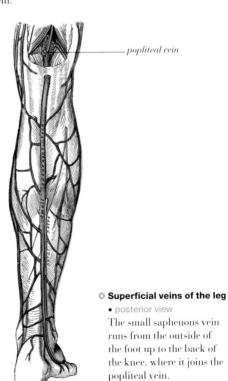

popliteal vein

Superficial veins of the leg
• posterior view

The small saphenous vein runs from the outside of the foot up to the back of the knee. where it joins the popliteal vein.

THE PORTAL VEIN

The portal vein in the abdomen is unique: it is the only part of the human system of blood vessels that does not lead either to or from the heart. Instead, it collects blood from the capillaries of all the major parts of the digestive system—from the stomach to the rectum—and other organs, including the pancreas and spleen *(see p. 194)*, and transports it to the liver. Here the portal vein splits again into many tiny capillaries, thus enabling the liver to carry out one of its major functions—filtering the bloodstream to remove unwanted substances. Unwanted, in this case, can be either a temporary or permanent state of affairs. If there is too much sugar in the blood, for example, the liver removes it and stores it as glycogen, to be released as glucose when the level in the bloodstream falls too low. Alcohol, on the other hand, is converted into fats for transport to the areas where the body stores fat, where it is then removed from the bloodstream. Cirrhosis of the liver, its failure as a result of excessive alcohol consumption, happens because the alcoholic's liver cannot disperse the fat for storage as fast as it is generated, and the excess builds up along the veins and arteries within the liver itself, choking the organ's blood supply and thus killing its cells.

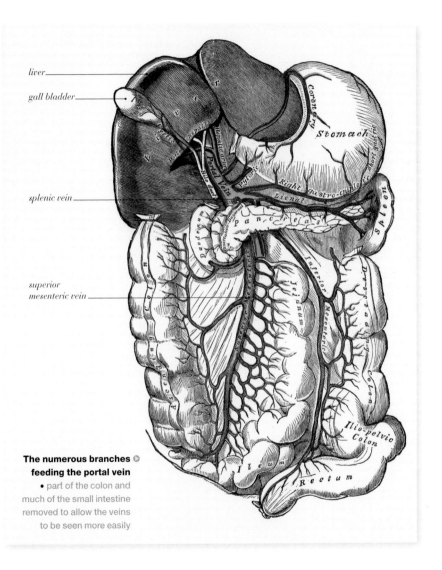

liver

gall bladder

splenic vein

superior
mesenteric vein

The numerous branches
feeding the portal vein
• part of the colon and
much of the small intestine
removed to allow the veins
to be seen more easily

THE SPLEEN

The spleen is made from extremely soft, stretchy tissue, and can act as a store for a modest amount of blood. Its main job, however, is to filter unwanted objects out of the blood. Inside its spongy tissue, the spleen contains an extremely large quantity of lymphocytes *(see p. 134)*, along with increasingly narrow blood vessels that branch into spaces containing a delicate web of connective tissue. This combination functions in much the same way as the lymph nodes *(see p. 198)*, but in this case the white blood cells are removing infections from the bloodstream rather than from the lymphatic system. The spleen is also responsible—along with the liver—for breaking down old red blood cells. Because of its extremely soft and delicate spongy nature, the spleen can easily be damaged by an object penetrating the lower part of the rib cage; an injury of this type leads to rapid loss of blood. Since the spleen is too soft to be easily repairable, the only medical solution is to remove it completely. However, the human body can in fact function perfectly well without it.

Cross-section through the spleen ⊙

This shows the branches of the splenic vein (the arteries have a similar arrangement) and the spongy structure of the organ's tissue.

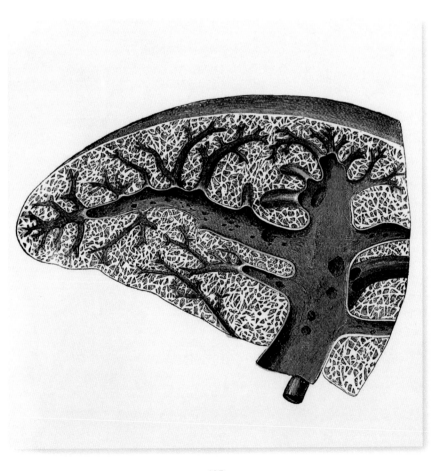

LYMPH GLANDS OF THE MAIN TORSO

Lymphatic fluid serves several purposes. First, it absorbs fat from the small intestine and distributes it to the rest of the body. This is necessary because, unlike sugar and protein, fat is not processed by the liver before being passed on to the rest of the body. Second, it forms a vital part of the immune system. The fluid also transports the remains of dead cells for reuse or disposal, and returns blood plasma lost from the capillaries to the bloodstream. The lymphatic system, like that for blood, has branches stretching into almost every part of the body. In every other way, however, the lymphatic system is simpler. There is no central pump like the heart; instead the lymphatic vessels have valves to ensure that fluid can only move in one direction, and the normal movement of muscles squeezes and expands the walls of the vessels, pushing fluid along.

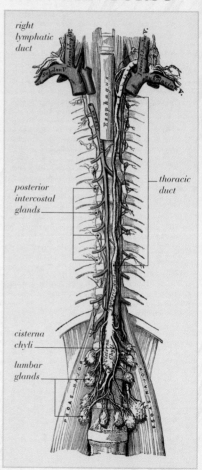

right lymphatic duct

posterior intercostal glands

thoracic duct

cisterna chyli

lumbar glands

Thoracic and right lymphatic ducts

The lymphatic system gathers fluid from numerous very small capillaries in all parts of the body, which steadily merge into larger vessels before, eventually, entering the bloodstream just above the heart. The two main trunks are the right lymphatic duct, which collects lymphatic fluid from the right arm and the right side of the head and neck, and the thoracic duct, which collects it from everywhere else in the body. These pass fluid into the right and left subclavian veins (see p. 141) respectively, where these are joined by the internal jugular veins.

◐ **Lymphatic vessels and glands in the male pelvis**

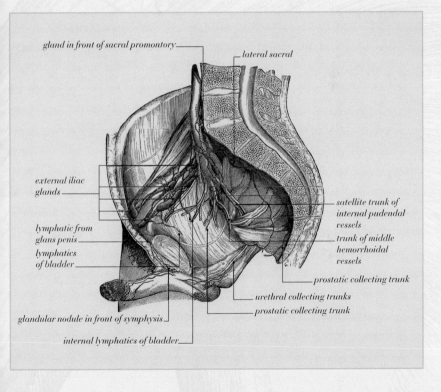

gland in front of sacral promontory

lateral sacral

external iliac glands

lymphatic from glans penis

lymphatics of bladder

satellite trunk of internal pudendal vessels

trunk of middle hemorrhoidal vessels

prostatic collecting trunk

urethral collecting trunks

prostatic collecting trunk

glandular nodule in front of symphysis

internal lymphatics of bladder

LYMPH GLANDS OF THE HEAD AND NECK

Lymph glands, also called lymph nodes, are found all over the body, but many of them are clustered together in certain areas. They are found particularly in the head and neck, in and around the pelvis, and around the shoulder

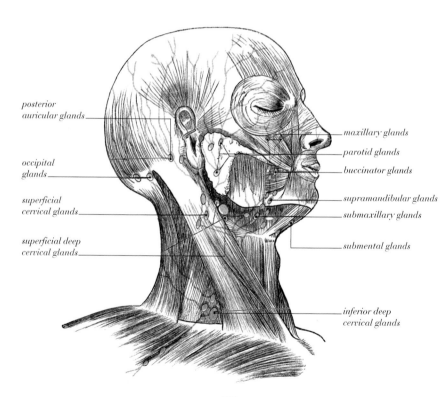

posterior auricular glands

occipital glands

superficial cervical glands

superficial deep cervical glands

maxillary glands

parotid glands

buccinator glands

supramandibular glands

submaxillary glands

submental glands

inferior deep cervical glands

and armpit. These glands are effectively filters in the lymphatic system, and contain large numbers of lymphocytes *(see p. 134)* which, along with a network of connective tissue inside the gland, work to destroy bacteria and viruses.

⊙ **Network of shallow lymphatic vessels and lymph nodes around the head and neck**
While lymph nodes exist in large numbers around the lower part of the head, there are normally none above ear level. There are several under the tongue and around the lower jaw but, again, none higher up the face.

Deeper lymphatic vessels ⊙ and glands in the head, neck, and shoulders
• clavicle bone removed
Pectorals major and minor divided to show cluster of lymph glands around axillary vein and artery.

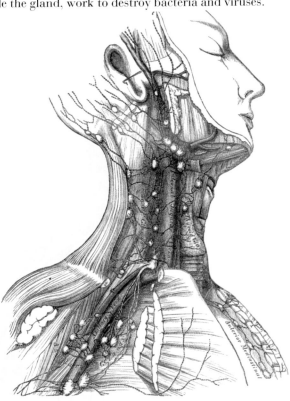

LYMPH GLANDS OF THE LIMBS

The lymphatic system in the arms and legs can be divided into two parts: deep and superficial. The deep lymphatic vessels, on which most of the lymph nodes are found, run adjacent to the major blood vessels. The lymph nodes that accompany these are mostly found near the joints, and become both larger and more numerous moving inward from the hand or foot toward the torso. The superficial lymphatic vessels run immediately underneath the skin. In the arm, these are typically accompanied by only one or two lymph nodes at each elbow and shoulder, with those in the armpit being much larger than those of the crook of the elbow. The leg has no superficial lymph nodes at all, except for the superficial inguinal nodes near the hip, but these are both larger and more numerous than the superficial lymphatic nodes of the arm.

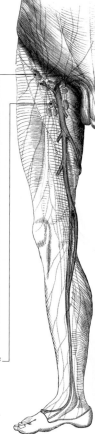

superficial inguinal nodes

superficial subinguinal nodes

**Superficial lymphatic system ◉
of the leg and abdomen**
Like the veins, the lymph
vessels have valves to prevent
fluid from flowing backward
toward the toes.

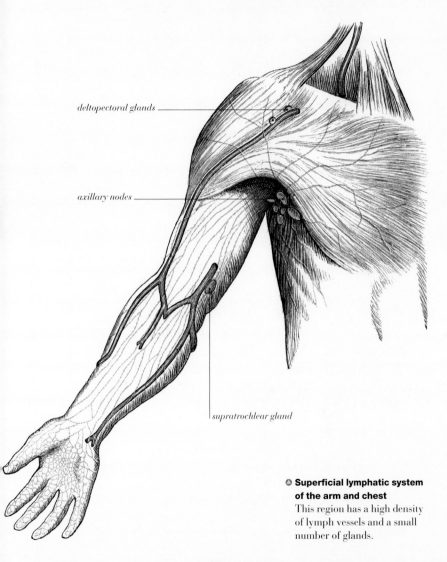

deltopectoral glands

axillary nodes

supratrochlear gland

**Superficial lymphatic system
of the arm and chest**
This region has a high density
of lymph vessels and a small
number of glands.

TYPES OF NEURON

The human nervous system contains two main classes of cell, each of which is subdivided into numerous types: nerve cells (or neurons) and glial cells, sometimes called neuroglia. The glial cells provide all the support functions that the neurons require to function properly—including transporting nutrients, producing myelin (a fatty chemical that forms a protective channel around the nerve fibers), and even actual physical support.

Glial brain cells of ○
two different types
Although glial cells do not produce neurotransmitters (axons), it is now believed that some of them may help in passing signals from one neuron to another. At right, A is a cell with branched processes, while B is a spider cell with unbranched processes.

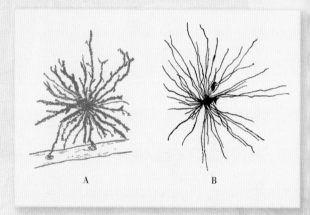

A   B

axon

cell body

cell nucleus

dendrite

Different nerve cells ▹

Although different types of neuron can have very dissimilar appearances, the basic structure is the same in each. Every nerve cell consists of a main body (the soma) which contains the cell nucleus, and two or more long, thin projections. One of these projections is the axon. It carries the electrical signal out of the nerve cell to one or more destinations. All of the others are the dendrites, which bring information into the cell. In the human body, the longest nerve cells are those running from the lumbar region of the spine to the toes; they can be over a meter in length. The axon, protected by myelin, forms almost the entire length of these cells.

THE AUTONOMIC NERVOUS SYSTEM

The human body performs a number of activities automatically, without any need for conscious effort. These include breathing, beating of the heart, peristalsis in the digestive system, and other functions that need to be carried out continuously, without risk of interruption when conscious thought is concentrated elsewhere. These activities are controlled by the autonomic nervous system, which is divided into two subsections: the sympathetic and parasympathetic nervous systems. Generally speaking, the two systems work together to ensure that energy and swift responses are available when required (the sympathetic system) but resources are not wasted keeping the body on high alert unnecessarily (the parasympathetic system). Control of the digestive system is sometimes considered to be handled by an independent system, called the enteric, since it is possible for this to function even when it is entirely cut off from the rest of the nervous system. The cranial, spinal, and autonomic nerves and their ganglia are collectively known as the peripheral nervous system.

Autonomic nervous system

The main parts of the system are the chain of ganglia, which runs down the vertebral column, and several large concentrations of nerves called plexuses, farther forward in the body. Each ganglion is a nerve-cell cluster that controls a particular set of nerves. The plexuses, like the ganglia, are connected to each other, to the other parts of the autonomic nervous system, and also to the central nervous system *(see p. 206)*. The most important of these are generally thought to be the cardiac plexus, which is next to the heart, the solar plexus, which is behind the stomach, and the hypogastric plexus, which is within the pelvis.

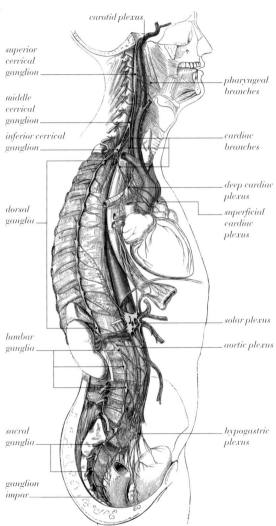

carotid plexus

superior cervical ganglion

middle cervical ganglion

inferior cervical ganglion

dorsal ganglia

lumbar ganglia

sacral ganglia

ganglion impar

pharyngeal branches

cardiac branches

deep cardiac plexus

superficial cardiac plexus

solar plexus

aortic plexus

hypogastric plexus

BASE OF THE BRAIN

The human central nervous system, comprising the brain and spinal cord, is much larger and more complex than the autonomic systems and controls both deliberate and reflex movements. A reflex movement is not the same as an involuntary one: the beating of the heart is an involuntary movement; pulling your hand away from a hot object is a reflex. The autonomic nervous system *(see p. 204)* consists of numerous small, interconnected clusters of nerves, at least some of which are believed to be able to continue working without any connection to the rest of the nervous system. The central nervous system, however, receives information from the nerves in the rest of the body (the peripheral nervous system, which includes the autonomic system), makes decisions, and then passes orders based on those decisions back to the peripheral system. If the brain is cut off from all or part of the rest of the nervous system, the results are severe: loss of sensation, paralysis, or even death. The brain itself is able to work around relatively minor damage to an impressive extent; even so, damage such as that caused by a stroke or a serious head injury is frequently fatal.

The human brain ⊙

• inferior view

The twelve pairs of major nerves to and from the brain are shown in yellow. These are called the cranial nerves and are numbered (always in Roman numerals) from the olfactory nerves at the front toward the back.

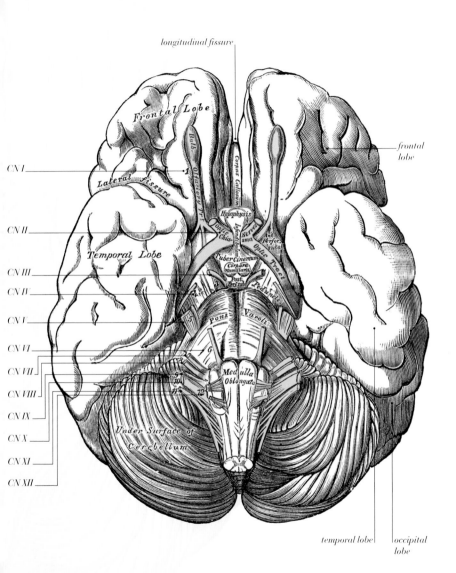

longitudinal fissure

Frontal Lobe

frontal lobe

CN I

Lateral Fissure

Bulb Olfactory Tr.

Corpus Callosum

Temporal Lobe

Hypophysis

CN II

Tractus Opticus

Optic Chia.

Nerve Infundibulum

Ant. Perfor. subst.

Ant. Perfor. subst.

Tuber Cinereum

Corpora Mamillaria

CN III

Post. Perforat.

CN IV

Peduncle

CN V

Pons Varolii

CN VI

9

Medulla Oblongata

CN VII

CN VIII

10

11

12

CN IX

CN X

Under Surface of Cerebellum

CN XI

CN XII

temporal lobe

occipital lobe

THE BRAIN IN CROSS-SECTION

The brain consists of two distinct layers: the gray matter, which is formed of the bodies and unprotected, thread-like extensions (dendrites) of the nerve cells *(see p. 202)*, covers the surface of the folds; the interior is filled with white matter, which is formed of the myelin-covered axons that lead from the cells in the gray matter to connect them. Although it is a common conception that the brain is divided into the left and right hemispheres, which each perform different tasks, this is not entirely true. The two hemispheres are connected by a large bundle of nerve tissue called the corpus callosum, in addition to smaller, more specialized connections called commissures. Many areas have exact equivalents in the other hemisphere, and even those that are not duplicated still share information and, if damaged, may have their functions taken over by a structure on the other side of the brain.

Inner surface of 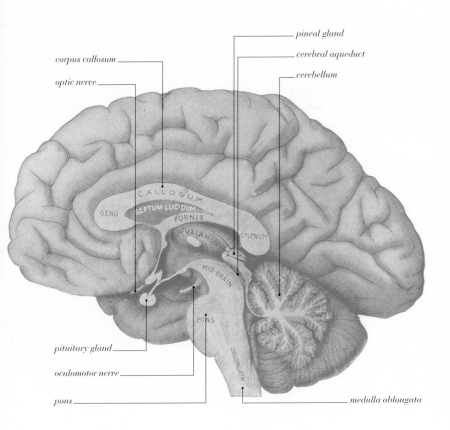
right hemisphere
of the brain

pineal gland

cerebral aqueduct

cerebellum

corpus callosum

optic nerve

CALLOSUM

GENU

SEPTUM LUCIDUM

FORNIX

THALAMUS

SPLENIUM

MID-BRAIN

PONS

OBLONGATA

pituitary gland

oculomotor nerve

pons

medulla oblongata

LATERAL VIEW OF THE BRAIN

The brain is smaller than the cavity of the skull into which it fits, and is surrounded by a "cushion" of fluid that helps to protect the soft fragile nerve tissue from the damage that might otherwise be caused by impact or by sudden movements—for example, whiplash—that could cause the brain to be thrown against the inside of the skull. This cerebrospinal fluid (CSF) also fills the ventricles that exist within the brain itself, supporting the brain from inside and preventing the soft structure from collapsing under its own weight.

**Outer surface ◉
of the left
hemisphere of
the brain**

• Different lobes
in different colors
The deep
crevices between
folds of brain
tissue are called
fissures.

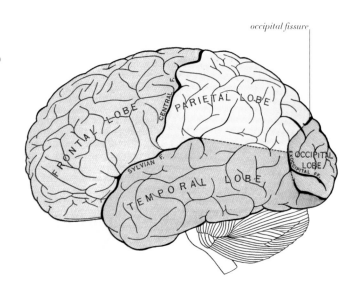

occipital fissure

Outer surface of the left hemisphere of the brain

The extremely complex arrangements of folds (gyri) and fissures (sulci) increase the area of the brain's surface, on which the gray matter is found, without changing its volume. This allows the brain to be fitted into a smaller, more conveniently shaped space than would otherwise be possible. The large surface area also allows heat to be removed more easily—an important feature because the brain uses a very large amount of energy (all of which is ultimately converted to heat), but is very sensitive to temperature and must not be allowed to get too hot.

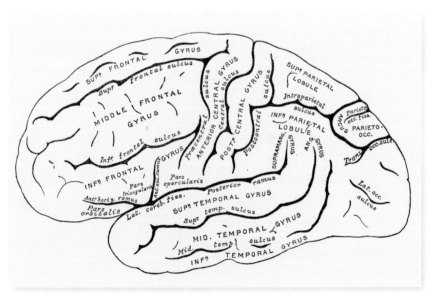

THE HINDBRAIN

The hindbrain lies at the lower part of the back of the skull, beneath the occipital and temporal lobes of the main hemispheres *(see p. 211)*. Definitions of which parts of the brain make up the hindbrain vary somewhat, but the three main structures are the medulla oblongata, pons, and cerebellum. The first two make up what is sometimes called the brain stem, which passes information between the top of the spinal cord, the cerebellum, and cerebral hemispheres. It is involved in controlling the automatic functions of the autonomic nervous system *(see p. 204)*.

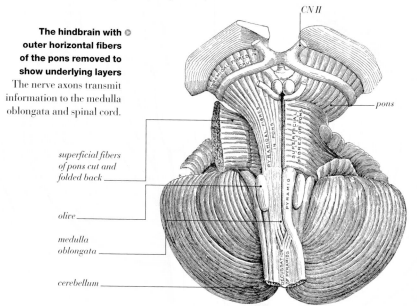

The hindbrain with ◉ outer horizontal fibers of the pons removed to show underlying layers
The nerve axons transmit information to the medulla oblongata and spinal cord.

CN II

pons

superficial fibers of pons cut and folded back

olive

medulla oblongata

cerebellum

The cerebellum is a larger and much more complex structure than the brain stem, though still much smaller than the cerebral hemisphere. It is closely involved in both the control of movement and the process of learning—particularly learning physical skills—and also appears to influence comprehension of language. Unlike the cerebral hemispheres, which each control the opposite side of the body, nerves in the cerebellum (itself divided into two hemispheres) generally affect muscles on the same side of the body.

The complex ⊙
under-surface
of the cerebellum
Although small when
compared with the two
cerebral hemispheres,
the cerebellum actually
contains a very large
number of neurons,
probably as many as
the two hemispheres
put together, and
possibly even more.

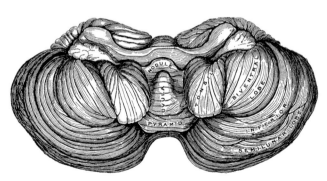

NERVES OF THE MOUTH, VOICE, RESPIRATION, AND DIGESTION

Of the 12 cranial nerve pairs *(see p. 206)*, the ninth to twelfth pairs contain, between them, almost all of the motor and sensory nerves related to the mouth, throat, lungs, and the digestive system as far as the middle part of the colon. The tenth cranial nerve, the vagus nerve, also connects to the parasympathetic nervous system *(see p. 204)* and has an effect on involuntary behavior including heartbeat, peristalsis, and sweating. This nerve is the most widely distributed of the cranial nerves, and consists of many

Distribution of the ninth ◉ (glossopharyngeal), tenth (vagus), and eleventh (accessory) cranial nerves on the left-hand side of the body
The eleventh cranial nerve divides rapidly into two sections: one of these controls the trapezius and sternomastoid muscles at the top of the back, while the other connects to the vagus nerve just after leaving the skull.

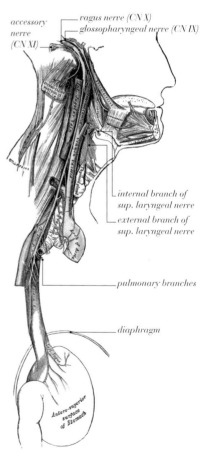

accessory nerve (CN XI)

vagus nerve (CN X)

glossopharyngeal nerve (CN IX)

internal branch of sup. laryngeal nerve

external branch of sup. laryngeal nerve

pulmonary branches

diaphragm

Antero-superior surface of Stomach

neurons working together, extending from the point where it leaves the brain stem down through the neck and chest into the upper part of the abdomen. The much smaller ninth and twelfth cranial nerves are connected to the mouth, with the twelfth nerve controlling the muscles of the tongue and several of those below the hyoid bone, while the ninth pair receives sensory information—both touch and taste—from the back of the mouth and the pharynx.

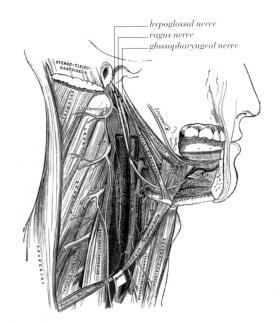

Deep dissection of neck and mouth to show the contents of the carotid sheath and the course of the hypoglossal nerve

hypoglossal nerve
vagus nerve
glossopharyngeal nerve

NERVES OF THE EYE AND ORBIT

Five of the twelve pairs of cranial nerves are involved with the eye and orbit. The second cranial nerve (or optic nerve) consists entirely of the axons of nerve cells behind the retina, carrying information to the optic centers of the brain. Although the number of fibers making up the optic nerve seems large (about one million), the retina contains over 100 million sensor cells (rods and cones). It has been suggested that the neurons at the back of the eye perform some of the necessary processing of the image before passing the information through the optic nerve to the brain. The fifth cranial nerve provides the sense of touch for various parts of the face, including the eyeball and surrounding area. The

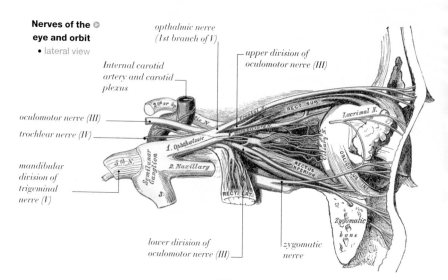

Nerves of the eye and orbit
- lateral view

opthalmic nerve (1st branch of V)

upper division of oculomotor nerve (III)

Internal carotid artery and carotid plexus

oculomotor nerve (III)

trochlear nerve (IV)

mandibular division of trigeminal nerve (V)

lower division of oculomotor nerve (III)

zygomatic nerve

third, fourth, and sixth cranial nerves, meanwhile, control the muscles of the eye *(see p. 82)*: the fourth is responsible for the superior oblique muscle, while the sixth controls the lateral recti muscles. All the other muscles connected to the eyeball, as well as the ciliary muscle and iris inside the eyeball, are controlled by the third cranial nerve, which also controls the muscles that hold the eyelid open.

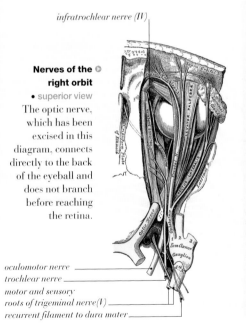

infratrochlear nerve (IV)

Nerves of the ◗
right orbit
• superior view
The optic nerve, which has been excised in this diagram, connects directly to the back of the eyeball and does not branch before reaching the retina.

oculomotor nerve ——————
trochlear nerve ——————
motor and sensory
roots of trigeminal nerve(V) ——————
recurrent filament to dura mater ——————

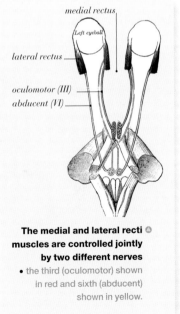

medial rectus

Left eyeball

lateral rectus ——————

oculomotor (III) ——————
abducent (VI) ——————

The medial and lateral recti ◗
muscles are controlled jointly
by two different nerves
• the third (oculomotor) shown in red and sixth (abducent) shown in yellow.

NERVES OF THE NOSE, FACE, AND JAW

Apart from the eyes *(see p. 216)*, nose, and mouth *(see p. 214)*, much of the rest of the face and jaw is served by the fifth cranial nerve (trigeminal). This transmits both sensory and motor information, although the sensory part of the nerve is much larger. Just inside the skull, the nerve forms the trigeminal ganglion, from which it separates into three branches. The uppermost of these provides the sense of touch for the area around the eye, while the lower two supply the nose and lower jaws respectively. These lower branches contain large numbers of both sensory and motor neurons, although information about smell is passed from the olfactory neurons in the nose to the first cranial nerve pair (the olfactory nerves), which are used solely for this purpose.

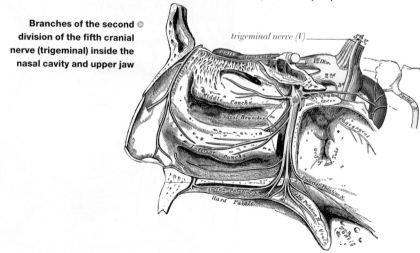

Branches of the second division of the fifth cranial nerve (trigeminal) inside the nasal cavity and upper jaw

trigeminal nerve (V)

Major branches of the second and third divisions of the fifth cranial nerve (trigeminal)

The upper of these two branches consists entirely of sensory nerves and provides the sense of touch for the area around the nose and upper jaw, as well as the nerves connected to the roof of the mouth and teeth of the upper jaw. The lowest of the three divisions contains both sensory nerves for the lower jaw and its teeth and for the outer ear, and motor neurons, which control the muscles of the cheeks and lower jaw that are used for chewing.

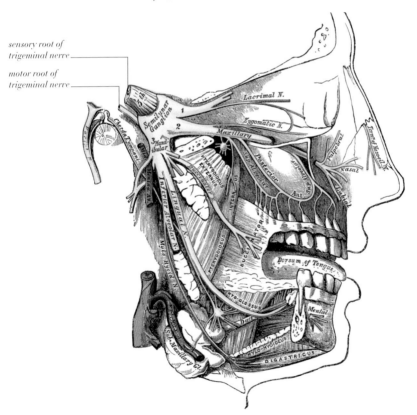

sensory root of trigeminal nerve

motor root of trigeminal nerve

NERVES OF THE SCALP, FACE, AND NECK

The movements of facial muscles—including the muscles on the surface of the skull around the ears—are controlled almost entirely by the facial nerve (the seventh pair of cranial nerves). This nerve, like the fifth cranial nerve, includes both motor and sensory neurons, with the motor neurons forming by far the greater part of the nerve in this case. Its numerous branches control the muscles at the front and back of the scalp, all of the muscles involved in facial expression, including those around the eyes, down the sides of the nose, and surrounding the mouth. The sensory branches of the seventh cranial nerve are responsible for transmitting information from most of the taste buds to the brain, and also for some of the sense of touch on the sides of the head, especially around the ear. Its final purpose is to provide the nervous system's link to the salivary glands, controlling the amount of saliva produced.

Arrangement of nerves on the ○ surface of the head and neck

While the sense of touch of the front of the head is supplied by the cranial nerves, those taking sensory information from the skin at the back of the scalp and in the neck are extensions from the cervical spinal nerves in the neck.

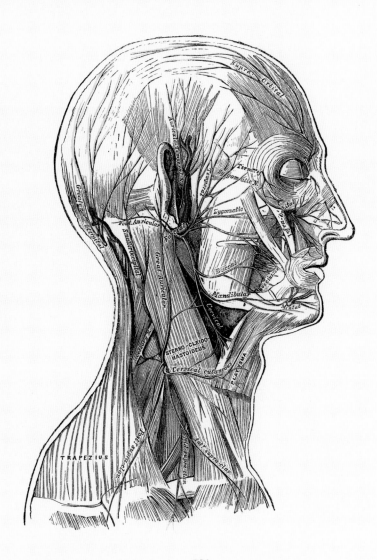

Supra-Orbital

Auricularis superior

Great occipital

Small occipital

Post. Auricular

Facial

Great Auricular

Temporal

Zygomatic N.

Zygomatic

Orbital

Temporal

Buccal

Masseter

Buccal

Nasal

Mandibular

Cervical

Mental

STERNO-CLEIDO-MASTOIDEUS

Cervical cutaneous

Platysma

TRAPEZIUS

Post. supraclav.

Lat. supraclav.

THE SPINAL CORD

The spinal cord, like the brain, consists of gray matter and white matter. However, the gray matter is in the center—forming an X- or H-shape in cross section, depending on its position—surrounded by white matter. The very center of the spinal cord is occupied by a small hollow space containing the cerebrospinal fluid *(see p. 210)*. The size of the spinal cord is not constant, nor does it shrink steadily from top to bottom, although it does generally become smaller as it descends the vertebral column. There are, however, two sections where it forms a bulge: one in the lower part of the neck where the nerves for the arms are attached; the other in the upper part of the lumbar region where it is joined by the nerves from the legs. The lower of these two enlargements is in fact the bottom of the spinal cord proper, which lengthens more slowly during childhood than does the vertebral column. Below this point, the space inside the vertebral column is filled with the nerves attached to the spinal cord.

Spinal cord in place with the vertebrae cut open around it

• posterior view

Generally speaking. the outer side of the spinal cord consists of nerves carrying sensory information to and from the brain. while the motor nerves. which control movement. are closer to the inner side.

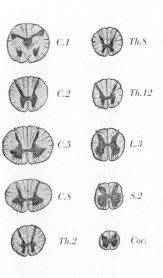

C.1   Th.8

C.2   Th.12

C.5   L.3

C.8   S.2

Th.2   Coc.

Sections through the spinal cord from the top of the spine downward

Upper cervical region (C.1, C.2); cervical enlargement (C.5, C.8); thoracic region (Th.2, Th.8); lumbar enlargement (Th.12, L.3); sacral region (S.2); and the final hollow filament of connective tissue (Coc.).

NERVES OF THE BACK OF HEAD AND TORSO

The muscles and skin of most of the body, including the back of the scalp, back and sides of the neck, and the entire torso from the shoulder downward are innervated by the spinal nerves. These in turn are connected to the spinal cord, which passes the information on to the brain for analysis and response.

Some reflex actions—for example, letting go of a hot object—are performed by the spinal cord responding directly to information from the peripheral nervous system, although that information is still passed to the brain in case an additional response is necessary. The spinal nerves—the name for those that are connected to the spinal cord rather than those that form it—all consist both of motor neurons, to control the muscles in the region to which they connect, and sensory neurons, to bring touch-related information back from the skin to the spinal cord and brain for processing.

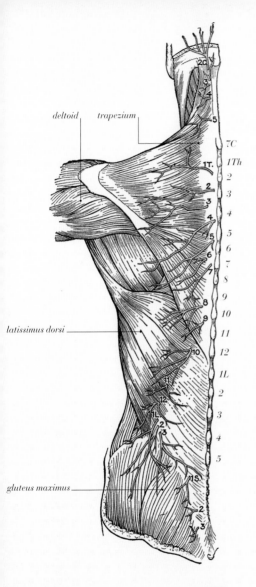

deltoid

trapezium

latissimus dorsi

gluteus maximus

2C
3
4
5
7C
1T. 1Th
2
3
4
5 5
6 6
7 7
8 8
9 9
10 10
11 11
12 12
1L 1L
2 2
3 3
4
5
1S.
2
3

225

◐ Arrangement of the nerves in the back of the head and back

These nerves are all branches from the spinal nerves. The head, neck, and other parts of the shoulder are supplied by the cervical (C) spinal nerves, whereas the main part of the back down to around the level of the waist are connected to the thoracic (Th) spinal nerves. Below the waist, parts of the body closer to the vertebral column are supplied by the sacral nerves, while those farther away are connected to the branches of the lumbar (L) nerves.

SUPERFICIAL NERVES OF THE ARM AND HAND

The nerves in the limbs are usually considered in two different groups. The superficial, or shallow, nerves of the arm or leg are those immediately beneath the skin. They transmit the sense of touch and related information, such as heat or cold, from the skin and hairs. They also contain the motor neurons controlling the tiny muscles that pull the individual hairs erect when we get goose bumps. The deep nerves of each limb consist of the major trunks (the bundles of nerves equivalent to the major arteries) and the clusters of motor neurons attached to the muscles to control the movement of the body. In comparison with the superficial nerves, these contain relatively few sensory neurons. In the arm, the superficial nerves—like the deep nerves—are all branches from the brachial plexus, which is formed by some of the lower roots of the cervical spinal nerves and upper thoracic spinal nerve (C.5–T.1).

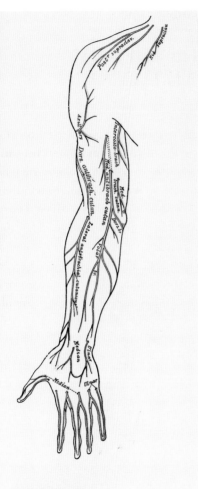

**Superficial nerves
of the right arm**

• anterior view.

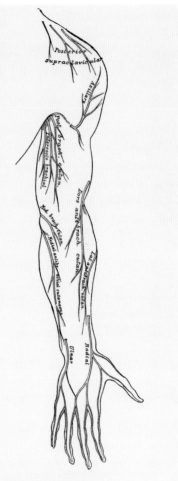

**Superficial branches of
the nerves of the right arm**

• posterior view.

DEEP NERVES OF THE ARM, SHOULDER, AND HAND

The deep nerves in the arm initially form three major trunks from the brachial plexus, which run down the neck and across the shoulder under the collarbone. Their many branches provide both motor and sensory neurons to the top and back of the shoulder and part of the chest, as well as the arm and hand. In common with all the other major plexuses of the peripheral nervous system, it connects to both the spinal cord and the autonomic nervous system.

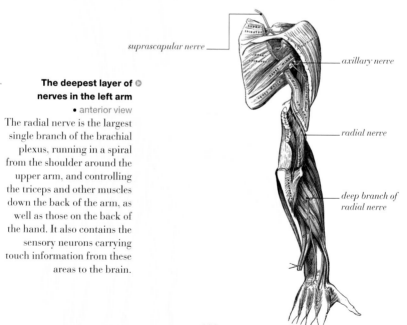

suprascapular nerve

axillary nerve

radial nerve

deep branch of radial nerve

The deepest layer of ◎ nerves in the left arm

• anterior view

The radial nerve is the largest single branch of the brachial plexus, running in a spiral from the shoulder around the upper arm, and controlling the triceps and other muscles down the back of the arm, as well as those on the back of the hand. It also contains the sensory neurons carrying touch information from these areas to the brain.

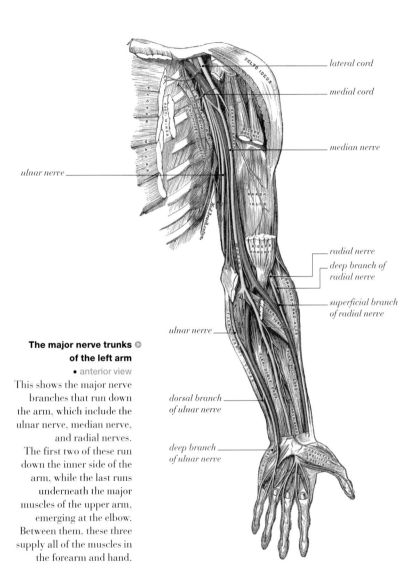

lateral cord

medial cord

median nerve

ulnar nerve

radial nerve

deep branch of
radial nerve

superficial branch
of radial nerve

ulnar nerve

dorsal branch
of ulnar nerve

deep branch
of ulnar nerve

The major nerve trunks ◉
of the left arm

• anterior view

This shows the major nerve
branches that run down
the arm, which include the
ulnar nerve, median nerve,
and radial nerves.
The first two of these run
down the inner side of the
arm, while the last runs
underneath the major
muscles of the upper arm,
emerging at the elbow.
Between them, these three
supply all of the muscles in
the forearm and hand.

NERVES OF THE LUMBAR REGION

Of the spinal nerves, the lumbar nerves are the longest and have the largest attachments to the spinal cord. Like the lower cervical spinal nerves, they form a large plexus, with numerous loops carrying information between the different lumbar nerves, as well as connections to the autonomic nervous system, particularly the parts controlling the digestive system. All the lumbar nerves originate within the vertebral column at about the level of the first lumbar vertebra (as do the sacral nerves). They then run down inside the vertebral column, each one emerging immediately beneath the vertebra for which it is named.

Lumbar and sacral nerves and their ⊙
branches attaching to various parts of the
lower back and surrounding area
The two uppermost branches from the lumbar plexus (the iliohypogastric nerve) control the muscles of the abdomen wall, in conjunction with the lower thoracic nerves. The third branch, the genitofemoral nerve, supplies the skin around the genitals and the top half of the inner thigh. The branches below this are responsible for sensation in the rest of the thigh and lower leg, as well as the control of many of the small muscles in the hip and front of the leg. The fifth and final lumbar nerve and the sacral nerves (see p. 232), form the sciatic nerve that runs down the back of the leg.

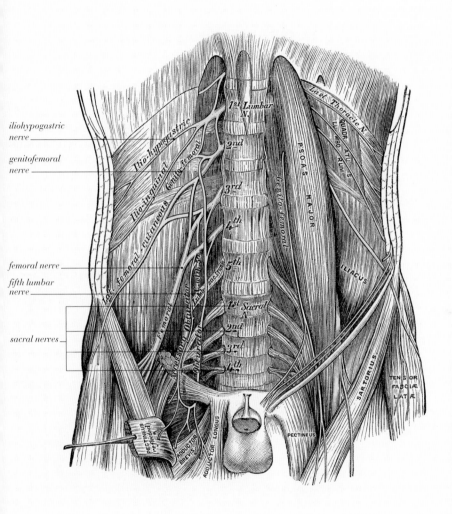

iliohypogastric
nerve

genitofemoral
nerve

femoral nerve

fifth lumbar
nerve

sacral nerves

THE SACRAL NERVES AND PELVIS

The sacral nerves, joined by the lowest branch of the lumbar plexus (p. 231), are responsible for most of the sense of touch and muscle control in the areas around the pelvis, particularly the buttocks, hips, and genitals, and also form the sciatic nerves that run past the hips and down the legs. Only a single small pair of nerves (coccygeal) emerges from the coccyx, surrounding the bone itself and the small muscle attached to its base. The sacral nerves, in addition to branches that control the gluteal muscles, their smaller companions around the pelvis and hips, and the muscles inside the pelvis, also control

Sacrum ◉

• posterior view

This shows the smaller posterior branches of the sacral nerves passing through the foramina (holes) in the sacrum to connect to the structures behind it. Because the spinal cord, to which they are attached, ends at about the level of the first lumbar vertebra, the sacral nerves often have a significant proportion of their length inside the vertebral column.

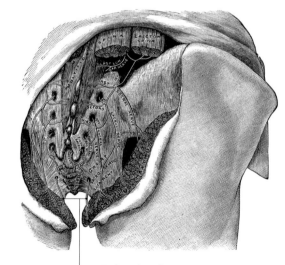

anterior branches of lower sacral nerves

the bladder and the various sphincters of the urinary and digestive systems. Slightly farther down, a branch of the sacral plexus—the pudendal nerve—provides the substantial channel for sensations from the genitalia, perineum, and surrounding area, including the anus. The pudendal nerve also contains motor neurons that control the erection of the penis and clitoris, as well as the muscular contractions of the vagina, cervix, and uterus or of the bands of muscle that move liquid down the urethra within the penis.

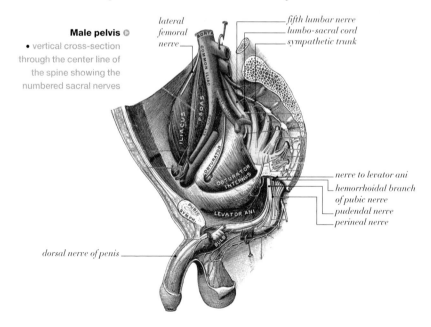

Male pelvis ⊙

• vertical cross-section through the center line of the spine showing the numbered sacral nerves

lateral femoral nerve

fifth lumbar nerve
lumbo-sacral cord
sympathetic trunk

nerve to levator ani
hemorrhoidal branch of pubic nerve
pudendal nerve
perineal nerve

dorsal nerve of penis

SUPERFICIAL NERVES OF THE LEG

The superficial nerves of the leg come from a slightly wider variety of sources than their equivalents in the arm. Those around the back and underside of the buttocks branch directly from the sacral nerves *(see p. 232)*. Much of the back and inside of the thigh is supplied by the posterior femoral cutaneous nerve, a branch from the sacral plexus.

Superficial nerve ○
branches of the right leg

• posterior view

Most of the superficial nerves on the back of the thigh are branches of the posterior femoral cutaneous nerve, while those on the back and sides of the calf are branches of the tibial nerve. The tibial nerve and common fibular nerves are divisions of the sciatic nerve.

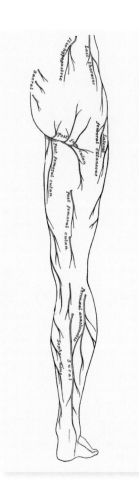

Superficial nerve ◎
branches of the right leg

● anterior view

The top parts of the front of the thigh, including the front of the inner thigh, are supplied by branches from the lumbar nerves, inside the pelvis *(see p. 230)*, while the skin on the rest of the front of the thigh receives branches from the femoral nerve, which also emerges from the lumbar plexus.

The figure labels read: Ilio-hypogast., Ilio-inguinal, Lumbo-inguinal, Lateral femoral cutaneous, Ant. femoral cutaneous, Int. of Com. peroneal, Saphenous, Superfl. peroneal.

DEEP NERVES OF THE LEG

The nerves controlling the muscles of the leg are all branches of one of two nerves. The first is the femoral nerve, which runs from the lumbar plexus over the front of the pelvis and down the front of the leg behind the rectus femoris. The second is the sciatic nerve which comes from the sacral plexus and runs down the back of the leg. It is the longest nerve in the body, extending unbroken, although not unbranching, from the sacral plexus to the sole of the foot. It is also the largest nerve, measuring around three-quarters of an inch in diameter near the top. Its branches contain the motor neurons for all of the muscles on the back of the leg from the hip down, for the joint of the hip, and for the sensory neurons for the skin of much of the lower leg and foot.

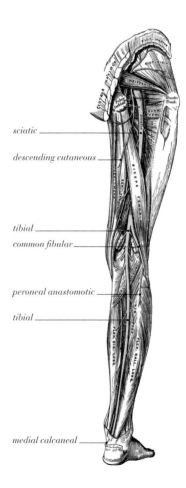

sciatic

descending cutaneous

tibial

common fibular

peroneal anastomotic

tibial

medial calcaneal

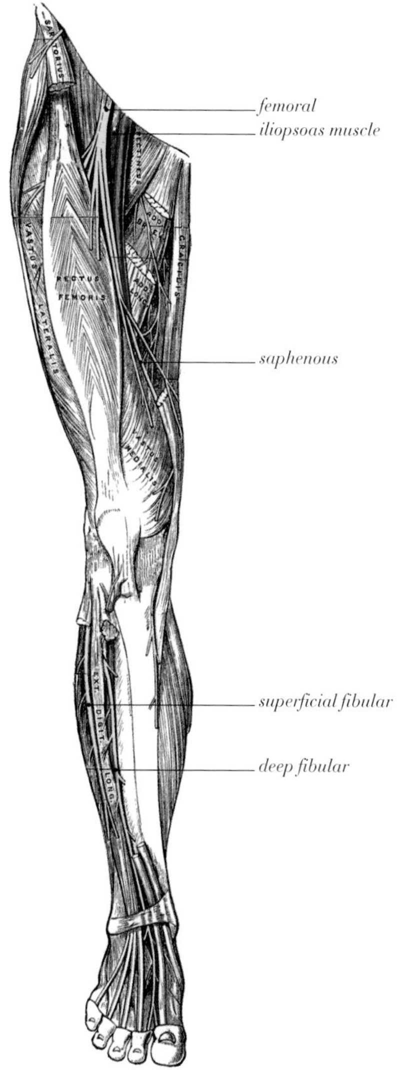

femoral

iliopsoas muscle

saphenous

superficial fibular

deep fibular

Deeper nerves of the right leg

• anterior view

The femoral nerve supplies the muscles of the quadriceps femoris and provides the sensation for the skin of much of the thigh.

Deeper nerves of the right leg

• posterior view

The sciatic nerve, running down the back of the leg, like the femoral nerve at the front, provides some branches to the hip joint. It then runs almost straight down the back of the leg to the knee, where it divides, with branches supplying all the muscles responsible for bending the leg.

NERVES OF THE HAND AND FOOT

As with the bones and blood supply, there is a degree of similarity between the arrangements of the nerves in the hands and feet. The nervous system of the hand is necessarily rather more complex than that of the foot.

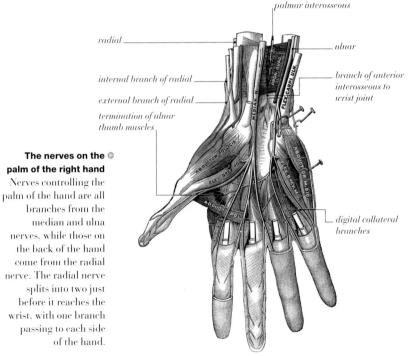

palmar interosseous

radial

ulnar

internal branch of radial

branch of anterior interosseous to wrist joint

external branch of radial

termination of ulnar thumb muscles

The nerves on the palm of the right hand
Nerves controlling the palm of the hand are all branches from the median and ulna nerves, while those on the back of the hand come from the radial nerve. The radial nerve splits into two just before it reaches the wrist, with one branch passing to each side of the hand.

digital collateral branches

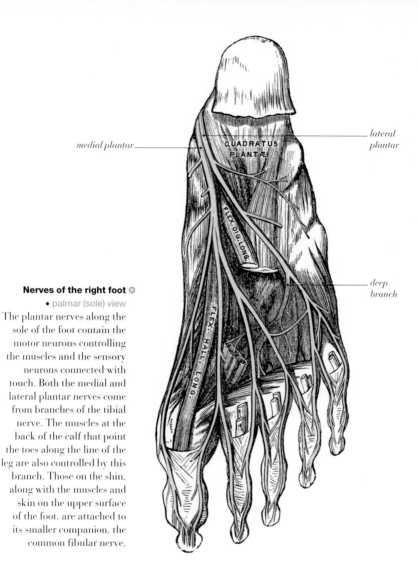

medial plantar

lateral plantar

QUADRATUS PLANTÆ

FLEX. DIG. LONG.

deep branch

FLEX. HALL. LONG.

Nerves of the right foot ◎

• palmar (sole) view

The plantar nerves along the sole of the foot contain the motor neurons controlling the muscles and the sensory neurons connected with touch. Both the medial and lateral plantar nerves come from branches of the tibial nerve. The muscles at the back of the calf that point the toes along the line of the leg are also controlled by this branch. Those on the shin. along with the muscles and skin on the upper surface of the foot, are attached to its smaller companion, the common fibular nerve.

THE EYE

The eyeball is an important but fairly fragile organ. It is protected in numerous different ways. Its position, set deep into the bones of the face, helps prevent damage from general impacts. The eyelids not only provide a physical covering to protect the eye from damage, but also contain a large number of small glands, or ducts, which continuously produce liquid to lubricate the movements of the eye and wash out dust or anything else that may fall onto the surface of the eye. Behind the eyelids is the conjunctiva, a

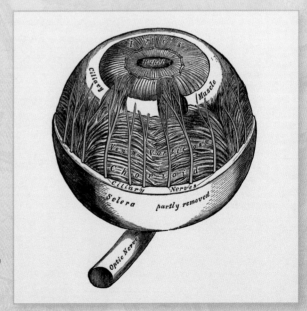

Eyeball ○
• parts of the sclera,
ciliary muscle,
and iris removed

membrane that covers both the posterior surface of the eyelids and the front of the eyeball. The outermost layer of the eyeball itself, the white of the eye, is called the sclera. It consists of strong white fibers, similar to those that make up ligaments and other connective tissue, and contains only a few small blood vessels. Inside this is the choroid, a brownish layer with numerous vessels carrying blood to the various parts of the eye, which is attached at the front to the ciliary muscle and the iris.

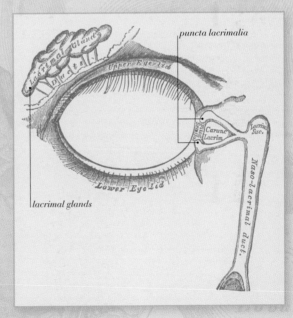

Arrangement of the various organs involved in producing tears around the right eye
The lacrimal glands, one for each eye, produce a small but steady flow of liquid similar to that produced by the glands in the eyelids, the amount of which can be increased on demand to remove unwanted substances like dust or smoke particles, or onion vapor, for example. Crying as a result of strong feelings produces a slightly more watery liquid.

INSIDE THE EYEBALL

The eyeballs are extremely complex. At the front of the eyeball, the bulge called the cornea is as strong as the sclera but is transparent. The cornea acts as a lens to focus light from a wide area through the pupil and into the eye. (The pupil is simply a hole in the middle of the iris, which is a mixture of circular and straight muscle fibers covered with a layer of unusual cells that result in its colors.) Behind the iris is the crystalline lens, which completes the job started by the cornea of focusing light from the outside onto the retina. The lens, although hard in the center, is softer and flexible at the edges, allowing its shape and with it the way in which light is focused within the eye to be changed by the ciliary muscle; however, the shape of the cornea is fixed. At the back of the eye, behind a thick transparent gel called the vitreous humor ("vitreous body" on the diagram), the many layers of the retina absorb light and convert it into electronic information, which is then transmitted as an image through the optic nerve to the brain.

Cross-section through the eyeball

There is a blind spot in each eye where the optic nerve and central artery pass through the retina, slightly to one side of the center. The space between the cornea and iris is filled with a liquid called aqueous humor which both protects the inside of the cornea and provides it with nutrients.

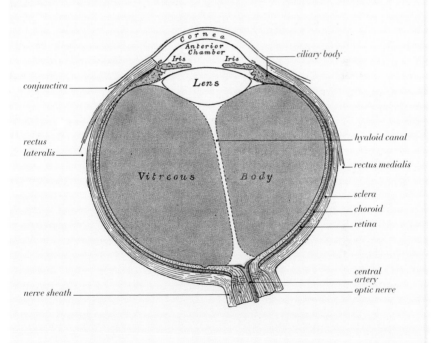

Cornea

Anterior Chamber

Iris — *Iris*

— *ciliary body*

conjunctiva ——

Lens

rectus lateralis ——

hyaloid canal

Vitreous — *Body*

rectus medialis

sclera

choroid

retina

central artery

optic nerve

nerve sheath ——

THE OUTER EAR

The external part of the ear (called the pinna or auricle) is designed to collect sound and channel it into the auditory canal, which then carries it to the inner parts of the ear. The pinna consists of a single large piece of cartilage, with a loose covering of fatty tissue and skin, although not all parts of the ear contain cartilage. Most notably, the earlobe at the bottom does not normally include any cartilage, which is why this is the easiest and safest part of the ear to pierce. In fact, cartilage takes longer to heal than skin and fascia—and punching a hole through it can cause splinters or cracks. There are also several muscles attached to the ear *(see p. 80)*. Although in humans these are small and serve little purpose, some animals have large, developed muscles which allow them actively to control the direction in which their ears are pointing.

The outer ear

Our ears are particularly good at channeling sound in the normal frequency range of human voices. The pinna transfers sound to the auditory canal (see p. 246) in a way that provides some information on the direction of the source of the sound, as well as the pitch and volume of the sound itself. This is how people are able to distinguish between sounds above and below or in front of and behind them. Which side (left or right) a sound is coming from can be distinguished by the brain on the basis of which ear received the sound first.

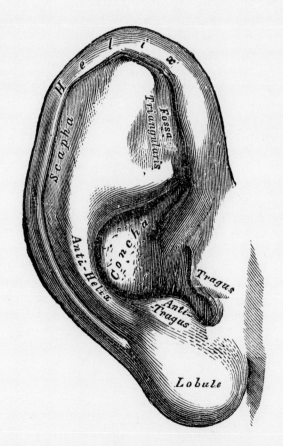

THE INNER AND MIDDLE EAR

The ear is divided into three parts: outer (parts visible on the outside), middle (eardrum and ossicles), and inner (labyrinth and cochlea). The middle ear receives sound from the outside as a vibration of the air in the auditory canal and of the eardrum, a membrane which closes off the inner end of the auditory canal. Attached to the inside of the eardrum is the first of three small bones, the ossicles; the first vibrates with the eardrum and causes the second to vibrate as well. This, in turn, vibrates the third ossicle.

A cross-section through the human ear

The eustachian tube, which is normally closed at the lower end, connects the middle ear to the top of the pharynx *(see p. 88)*. Opening it allows the pressure inside the eardrum to be changed to match the pressure of the air outside (this is what happens when your ears pop) and to prevent the eardrum from either bursting or being crushed by an excessive pressure difference.

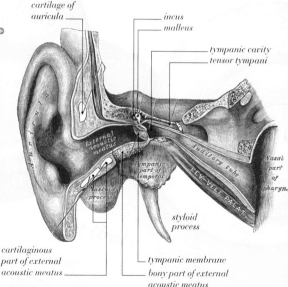

cartilage of auricula

incus

malleus

tympanic cavity

tensor tympani

External acoustic meatus

Auricula

Tympanic part of temporal

Mastoid process

Auditory tube

LEV. VELI PALAT.

Nasal part of pharynx

styloid process

cartilaginous part of external acoustic meatus

tympanic membrane

bony part of external acoustic meatus

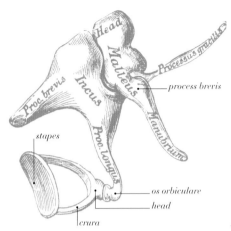

The third ossicle is attached to an opening in the membrane surrounding the inner ear and passes the sound inward. The inner ear consists of two parts which, although connected, serve different functions. Inside the spiral cochlea—which looks something like a snail shell—the sound vibrates in fluid-filled compartments, moving hairs which then produce an electrical nerve signal that is transmitted to the brain. The labyrinth provides our sense of balance, keeping us upright and helping our brain to compensate for movement.

The three small bones of the middle ear
The malleus (hammer) is closest to the eardrum, to which it is attached by the long thin spike called the manubrium. The flat part of the stapes (stirrup) covers the opening in the membrane of the inner ear, while the incus (anvil) connects the other two bones to each other.

The inner ear • with
the near side cut away
The fluid inside the labyrinth and cochlea transmits movement and vibration, via hairs, to the brain..

THE NOSE

The nose serves two purposes: it provides a passage for air in and out of the body when breathing and it contains a special tissue that provides us with our sense of smell. The part of the nose jutting from the face exists merely to protect the inner part: it consists of several pieces of cartilage attached to the facial bones. These form the outside of the nostrils, except for the septum of cartilage which separates the two nostrils, and they are surrounded by layers of surface tissue. The inside of the nostrils are lined with small hairs to trap dust before it gets too far into the

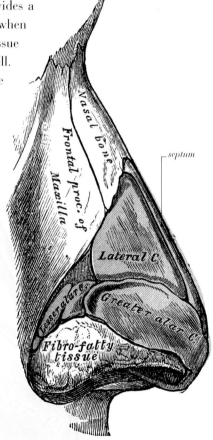

Nasal bone

Frontal proc. of Maxilla

septum

Lateral C.

Lesser alar

Greater alar C.

Fibro-fatty tissue

The nose (cartilages ◔ shown in blue)
• lateral view

body. Inside the nasal cavity in the skull *(see p. 20)*, about 3 inches (7.5 cm) from the nostrils, is the olfactory epithelium which provides the sense of smell. This contains many different types of olfactory neurons (each of which can detect a single specific smell molecule), basal cells (which can turn into any of the different types of olfactory neuron, to replace dying cells), and other cells (which produce mucus to protect the olfactory tissue). A typical human has around 40 million olfactory neurons, split between roughly 350 different types, but individuals do not always have exactly the same set of types, so two people will often be able to identify slightly different sets of smells.

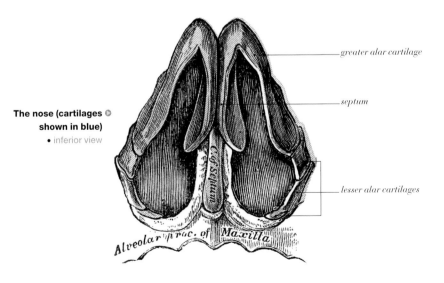

The nose (cartilages ◉ shown in blue)

• inferior view

greater alar cartilage

septum

lesser alar cartilages

THE TONGUE

This is the main organ of the sense of taste, and is important for speaking, and for eating. (Its muscular structure is discussed on *p. 86*.) Wrapped around the muscles is a mucous membrane of variable thickness. It is thicker on the upper surface than the underside, and thicker at the back than at the front. The mucus it produces lubricates the mouth and assists with swallowing and speech. Covering the surface from the tip to about two-thirds back are many

Upper surface ◉ of the tongue

The vallate or circumvallate papillae (tastebuds) are arranged in a V-shaped line separating the front two-thirds of the tongue containing the rest of the tastebuds from the posterior part. The fungiform papillae are just large enough to be seen in the image.

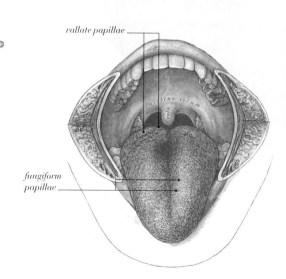

vallate papillae

palatine uvula

fauces

fungiform papillae

small bulges in this membrane. These are the papillae, or tastebuds. The back third of the tongue does not have any tastebuds, although the surface is bumpy owing to the many small lymph nodes it contains. Unlike the olfactory nerves *(see p. 249)*, the tastebuds are fairly general receptors identifying salt, sour, sweet, bitter, and savory flavors in varying degrees from different substances.

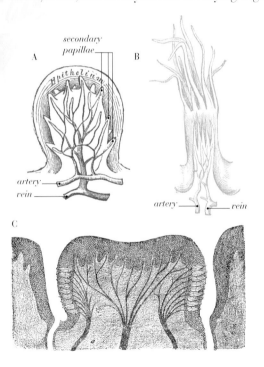

○ **Three of the four different types of tastebud**
The fungiform papillae (A) are mainly clustered at the front of the tongue. The long thin filiform papillae (B) are usually present in the highest number, scattered fairly evenly across the surface of the front two-thirds of the tongue, while the circumvallate papillae (C) are arranged in a row at the back. The foliate papillae (not shown) take the form of ridges across the tongue, and are mostly found close to the vallate papillae at the back.

THE SKIN AND HAIR

The skin contains and protects the rest of the body. It acts as a barrier against bacteria, limits water loss, absorbs UV light for the production of vitamin D, but at the same time contains pigments that prevent too much cancer-causing UV light from being absorbed. In addition, the tens of thousands of tiny nerve sensors in the skin's layers collect information about heat, cold, pressure and other sensations. Vibration or movement of the tiny hairs on the skin's surface also transmits sensations through the nerve endings that surround them.

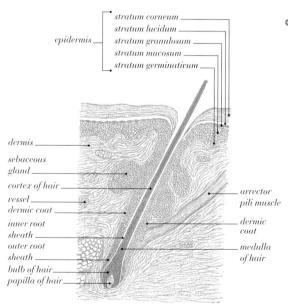

stratum corneum
stratum lucidum
epidermis — *stratum granulosum*
stratum mucosum
stratum germinativum

dermis

sebaceous gland

cortex of hair

vessel

dermic coat

inner root sheath

outer root sheath

bulb of hair

papilla of hair

arrector pili muscle

dermic coat

medulla of hair

◷ Skin with a hair follicle

• cross-section

Hairs are pulled erect by extremely small muscles called the arrectores pilorum, which connect a point near the base of the hair to the underside of the skin. The dermic coat around the hair generally contains a large number of fine nerve endings, which is why sensations provided by hair are often particularly strong. The hair follicle is the small pit in the skin in which the root of the hair sits, rather than the pale bulb at the very end of the hair, which is the new growth surrounding the rapidly dividing cells that form the hair as it grows.

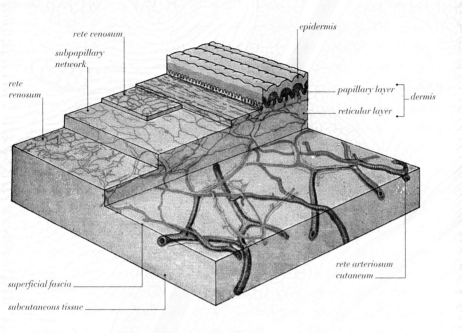

Labels on figure:
rete venosum
subpapillary network
rete venosum
epidermis
papillary layer
reticular layer
dermis
superficial fascia
subcutaneous tissue
rete arteriosum cutaneum

⊘ Arrangement of the many layers of the skin

The skin's layers are gathered into two groups called the epidermis and the dermis. The epidermis consists of layers of cells that contain a substance called keratin, a natural polymer which also makes up much of the structure of hair and nails, as well as the beaks, claws, feathers, and fur of most animals. These cells are produced by cell division at the bottom of the epidermis. As they move closer to the surface, the amount of keratin they contain increases, and they become harder and stiffen. Eventually the dead cells drop off the outermost layers or are rubbed off onto another object. The epidermis itself contains no blood vessels; it is supplied entirely by capillaries in the highest layers of the dermis beneath it. The dermis also contains many other structures like sebaceous glands (sweat glands) which exist in most areas of the skin, and the roots of hairs, as well as blood vessels, nerves, and so on. Beneath the dermis are the layers of superficial fascia (see p. 78).

THE CARTILAGE OF THE LARYNX: BREATHING

The larynx (or voice box) consists of nine pieces of cartilage: the epiglottis, thyroid, arytenoid (two), cuneiform (two), corniculate (two), and cricoid cartilage (the only complete cartilaginous ring in our airway), and several associated muscles and ligaments at the top of the trachea. These are immediately beneath the back of the tongue. They serve two main purposes: they help to protect the top part of the trachea from external damage and from misdirected food; and they are responsible for much of the complex control of sound required for speech. When a person is breathing, the muscles hold the cartilage in place so that air can flow without restriction between the trachea below and the nasal passages above.

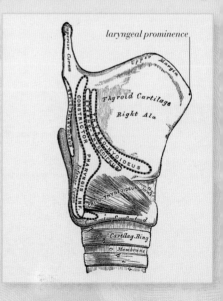

laryngeal prominence

Outside of the male larynx ○
• lateral view from right
The size and shape of the larynx controls the pitch of the voice to some extent. The thyroid cartilage is much larger in adult men than in women or children, resulting in the much more prominent male laryngeal prominence or "Adam's apple."

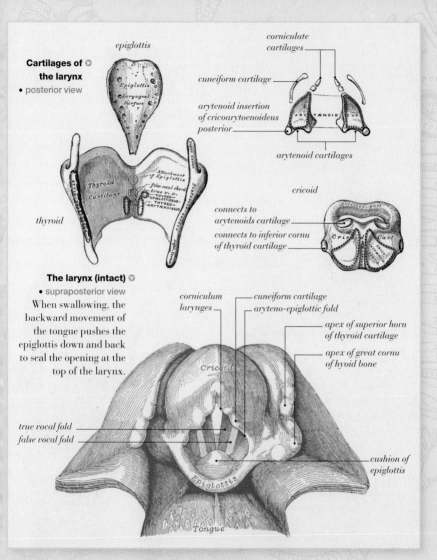

Cartilages of the larynx
- posterior view

epiglottis

corniculate cartilages

cuneiform cartilage

arytenoid insertion of cricoartoenoideus posterior

arytenoid cartilages

thyroid

cricoid

connects to arytenoids cartilage

connects to inferior cornu of thyroid cartilage

The larynx (intact)
- supraposterior view

When swallowing, the backward movement of the tongue pushes the epiglottis down and back to seal the opening at the top of the larynx.

corniculum larynges

cuneiform cartilage

aryteno-epiglottic fold

apex of superior horn of thyroid cartilage

apex of great cornu of hyoid bone

true vocal fold

false vocal fold

cushion of epiglottis

THE MUSCLES OF THE LARYNX: SPEAKING

The larynx, like the pharynx above it and the trachea below, has a thin lining of mucous membrane to ensure that it remains moist (a dry scratchy voice is precisely that). It is otherwise essentially a structure made entirely of cartilage. Inside the larynx, however, are a number of small muscles and four unusual and important narrow ligaments surrounded by a layer of mucous membrane. These form the vocal folds. In fact, the upper pair (called the false vocal folds, *see p. 255*) are not involved in making sound at all, but serve to protect the true vocal folds beneath them.

Larynx ⊙

• lateral view from right, half the thyroid cartilage removed to show the arrangement of muscles
These muscles can be divided into two groups: those moving the vocal folds in and out of the airstream; and those controlling the tension of the vocal folds. In the first group, the lateral cricoarytenoids (one on each side) and the transverse arytenoid (a single muscle crossing the back of the larynx) bring the true vocal folds together, thus closing the rimaglottis (the gap between them). The posterior cricoarytenoid opens it again.

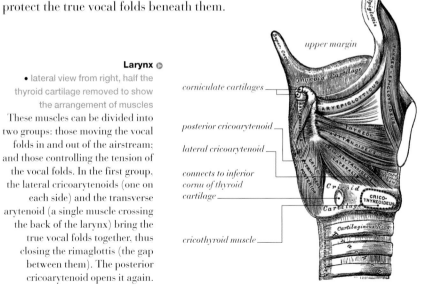

upper margin

corniculate cartilages

posterior cricoarytenoid

lateral cricoarytenoid

connects to inferior cornu of thyroid cartilage

cricothyroid muscle

256

The vocal folds vibrate when they are moved into the stream of air passing through the larynx, producing sound of different pitches depending on how tightly stretched they are. Between the changes in pitch and volume produced here, and the varying ways in which sound can be affected by movements of the parts of the mouth, an amazing range of sounds can be produced—no single human language uses all of them. The uvula—the piece of muscle in the back of the mouth shown vibrating when cartoon people are shouting or singing—has nothing to do with speech, but is for sealing the nasal passages when swallowing.

Interior of the larynx ◉
• view from above, epiglottis removed
Each vocal fold is controlled by two muscles: the cricothyroid, which tilts the thyroid cartilage, stretching the vocal fold; and the thyroarytenoid, which pulls the arytenoid cartilage inward, shortening the vocal fold.

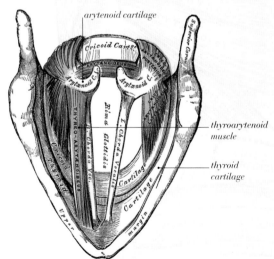

arytenoid cartilage

thyroarytenoid muscle

thyroid cartilage

THE TRACHEA AND BRONCHIAL TUBES

The trachea (or windpipe) is a roughly cylindrical tube that carries air down the throat from the bottom of the larynx into the chest, where it splits into two smaller passages called the bronchi. The left bronchus is both longer and narrower than the right because it must bypass the heart to reach the left lung. These bronchi run an inch or two (3–5 cm) farther down the chest before branching and entering the lungs, where their main branches rapidly subdivide into numerous small tubes.

The structure of the trachea is controlled by two contradictory requirements: it must be as flexible as possible to avoid restricting the movement of the neck, but it must also be as strong as possible, since serious damage, whether by crushing or piercing, could well be fatal. To achieve this, the trachea has alternating bands of cartilage and muscle (typically 15 or 20 of each in the trachea, with the pattern continuing down the main bronchi), with a membrane of strong elastic fibers down both the inside and outside of the tube. There are also a small number of muscle fibers running the length of the trachea down the outside of this membrane. Within the inner elastic membrane is a further layer, which uses a combination of tiny hair-like projections from individual cells and mucous to trap bacteria and small particles before they can reach the lungs and cause infection or other problems.

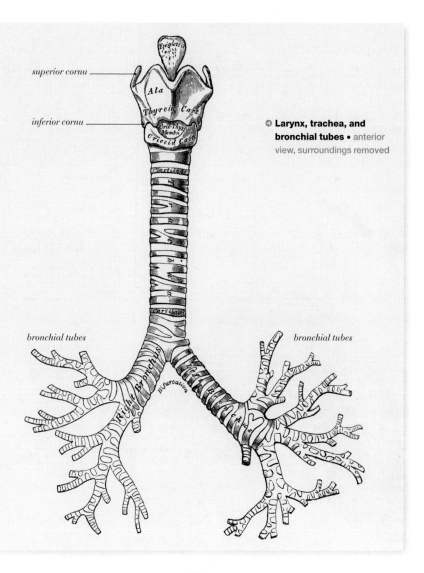

superior cornu

inferior cornu

Epiglottis

Ala

Thyroid Cart.

Crico-Thyr. Mem'bn.

Cricoid Cart.

Cartilage

Cartilage

bronchial tubes

bronchial tubes

Right Bronchus

Bifurcation

Left Bronchus

● **Larynx, trachea, and bronchial tubes** • anterior view, surroundings removed

THE LUNGS

The last and most important part of the respiratory system is comprised of the two lungs, which remove carbon dioxide from the bloodstream, replacing it with oxygen to be taken to the rest of the body. In a typical adult human, the lungs have the ability to hold around 10½ pints (5 liters) of air, of which around half will be expelled and replaced with each breath during normal breathing. The lungs are made of a very soft, spongy material, which is easily damaged by even the lightest pressure, hence the presence of the rib cage to prevent anything outside the body from crushing the lungs. The physical

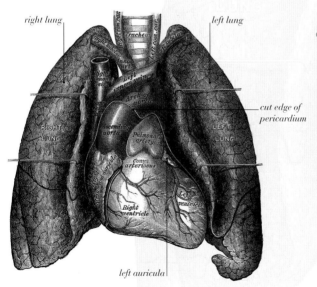

right lung

left lung

cut edge of pericardium

left auricula

◉ **The heart and lungs**
• anterior view
The right lung is shorter and wider than the left, and has a slightly greater volume overall (about 10%) as a result of the space occupied by the heart on the left-hand side of the chest.

Chest ◉
• cross-section, immediately above the top of the heart

fragility of the lungs is a result of their structure: a complex honeycomb of tiny spaces created by the interweaving of the minute end branches of the bronchial tubes *(see p. 258)* with the capillaries bringing blood to be supplied with oxygen. As part of their protection, the lungs are surrounded by membranes (the pleurae) that effectively attach them to the inside of the rib cage to prevent collapse. However, these membranes, which consist of two thin layers separated by a small amount of liquid, are themselves fragile, and damage to them can cause serious breathing problems.

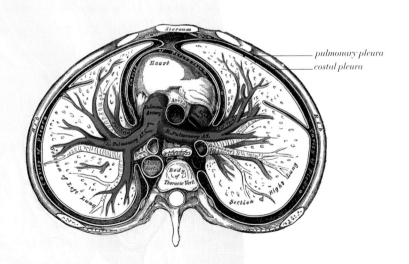

pulmonary pleura
costal pleura

THE TEETH

Digestive fluids like saliva and stomach acid can only affect the surface of food. The first stage of digesting any solid food is therefore to break it into smaller pieces and increase its surface area using the teeth. Humans will normally have two sets of teeth in their lifetime. The first set (often called the milk teeth) are smaller and fewer in number than the permanent teeth. Milk teeth begin forming in the unborn fetus, although they do not normally emerge through the skin of the jaws until between six months and two years after birth. Both sets take the same basic form, however, with a strong layer of dentine (a substance similar to ivory) protecting the soft and sensitive dental pulp in the center. The dentine itself is surrounded by a cement which binds it to the jaw below the gum, and by very hard tooth enamel above the gum.

Permanent teeth

• right-hand side of the jaw
These replace the milk teeth over a period of years, typically between the ages of about six and twelve years, as a child's jaws grow toward their adult size; there are 32 permanent teeth compared with only 20 milk ones.

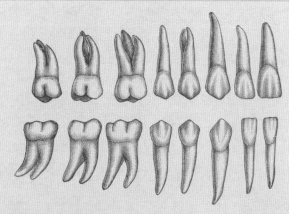

Half-profile view of jaws and teeth

The different types of teeth are well adapted for different types of food. Those at the front are designed for cutting, while the molars and bicuspids farther back can be ground together with sufficient force to break down the structure of most plant or animal foods.

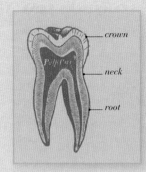

crown

neck

root

Molar • cross-section showing the enamel (white), dentine (gray), and central cavity containing the dental pulp (black).

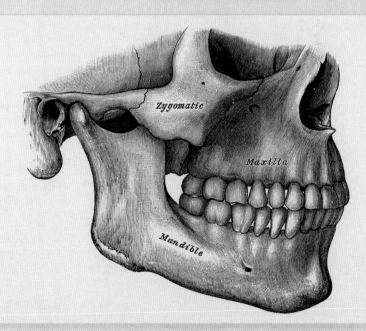

Zygomatic

Maxilla

Mandible

THE SALIVARY GLANDS

Saliva serves many purposes: it keeps the inside of the mouth moist; it contains enzymes that begin the process of digestion, especially of starches; it helps transport chewed food to the throat when swallowing; and it also contains natural antibacterial agents, which may help to prevent infections in the mouth and digestive system. Human saliva has not been shown to have any particular healing attributes, but the saliva of some animals has properties that can greatly increase the speed of healing. Saliva is very slightly alkaline, since acids damage tooth enamel, and it contains a number of chemicals that help to maintain the surface of the teeth.

Salivary glands ⊙

There are three main salivary glands on each side of the face and many smaller ones around the mouth. The parotid gland, which is the largest of the glands, passes saliva to the mouth through a duct that emerges from the cheek roughly level with the second molar tooth of the upper jaw. The sublingual and submandibular glands, beneath the tongue and just inside the back of the jawbone, respectively, are connected to ducts that introduce saliva into the mouth along the sides of the root of the tongue. The largest duct is at the front of the mouth, just beneath the tip of the tongue.

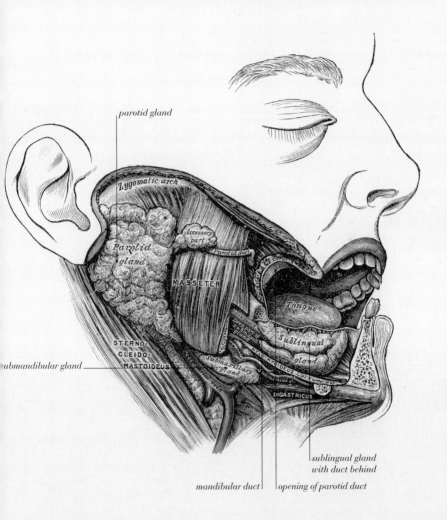

parotid gland

zygomatic arch

Accessory part

Parotid duct

Parotid gland

MASSETER

Tongue

Duct of submandibular gland

Sublingual gland

STERNO-CLEIDO-MASTOIDEUS

MYLO-HYOIDEUS

Submaxillary gland

submandibular gland

DIGASTRICUS

sublingual gland with duct behind

mandibular duct   opening of parotid duct

ESOPHAGUS: SWALLOWING

When food has been broken down into sufficiently small pieces, the muscles of the tongue, which have been lubricated with saliva, move the food to the back of the mouth *(see p. 86)*. At this point, the muscles of the pharynx *(see p. 88)* take over and move it farther back and down into the esophagus, making sure that it does not fall into the trachea instead. The esophagus, like the pharynx, is a muscular tube, arranged so that rings of alternately relaxed and contracted muscle will move slowly down it, pushing food toward the stomach in a series of wave-like movements (peristalsis). Both the esophagus and the pharynx are lined with large cells that are designed to produce a slippery mucus that helps the food to slide down. At the bottom of the esophagus is a separate ring of muscle called the cardia (sometimes called the cardiac valve, although it is not strictly a valve). This permits food pushed down the esophagus to enter the stomach and usually prevents anything from passing back up. When food does move back up, the result is acid indigestion, or heartburn.

Nose, mouth, pharynx, and larynx ◉
• cross-section

In humans, the passages for air and food effectively cross one another. While this arrangement of the pharynx makes choking much more likely than for many animals, it does have two advantages. First, it makes speech possible by vastly increasing the range of sounds we can produce. Second, it allows us to breathe through our mouths as well as our noses.

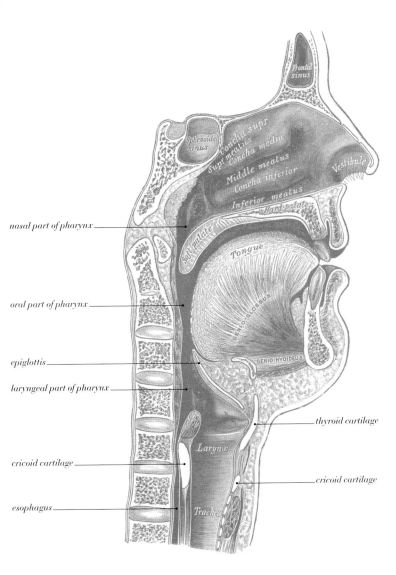

Frontal sinus

Sphenoidal sinus

Concha supr

Supr meatus

Concha média

Middle meatus

Concha inferior

Inferior meatus

Vestibule

Hard palate

Soft palate

nasal part of pharynx

Tongue

GENIOGLOSSUS

oral part of pharynx

GENIO-HYOIDEUS

epiglottis

laryngeal part of pharynx

thyroid cartilage

Larynx

cricoid cartilage

cricoid cartilage

esophagus

Trachea

THE ABDOMEN

The stomach and the 25 feet (7.5 meters) or so of intestine possessed by a typical adult human are found in the abdomen, along with many other organs, some of which, such as the liver, contribute to the digestive process without being part of the alimentary canal. The entire contents of the abdomen— between the pelvis at the bottom and the diaphragm at the top—are enclosed by the peritoneum, a single large membrane. This provides much less direct protection than either bone or muscle, but consists of two layers separated by a small amount of liquid (2 fl. oz/50 ml), allowing them and the organs contained within to move freely over each other, sliding the fragile digestive organs out of the way of damage.

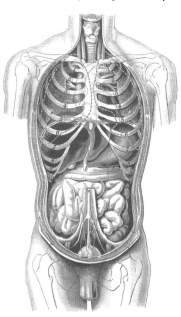

**General arrangement of organs in ◉
the torso and abdomen**

• the dotted lines show the
outline of the lungs

The diaphragm and the peritoneum beneath it come as high as the top of the fifth rib over the liver, which rises slightly farther into the chest than the stomach on the other side. The liver is normally on the right-hand side of the body, with the stomach on the left. A complete left–right reversal of the contents of the abdomen, while relatively rare, is neither unheard of nor in any way harmful.

The diaphragm rises much higher in
the center than might be expected: the
diagram accurately records its position.

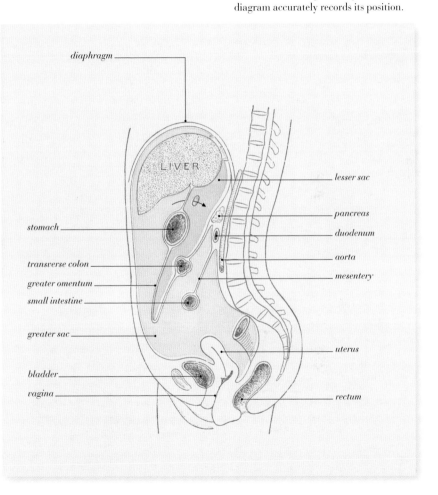

diaphragm

LIVER

lesser sac

pancreas

stomach

duodenum

transverse colon

aorta

greater omentum

mesentery

small intestine

greater sac

uterus

bladder

vagina

rectum

269

THE STOMACH

Once food enters the stomach, it is attacked with a potent mixture of hydrochloric acid and digestive enzymes, produced by cells in many small pits lining the internal surface. Since the enzymes in the stomach are mostly involved in breaking down the proteins of which muscle is made, other cells in these pits produce mucus to protect the stomach lining from its own digestive fluids. Hydrochloric acid will break down almost anything, given time, and so this protective mucus is continually replaced. While the rest of the digestive system has muscular walls with two layers, the stomach has three. The two outer layers have the same arrangement of muscle fibers as other parts of the digestive tract. The innermost muscular layer is unique, with fibers running diagonally around the curvature of the stomach.

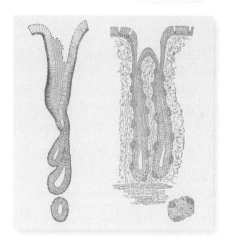

○ **Pits in the stomach lining** • close detail Enzymes, acid, and mucus are produced in these pits.

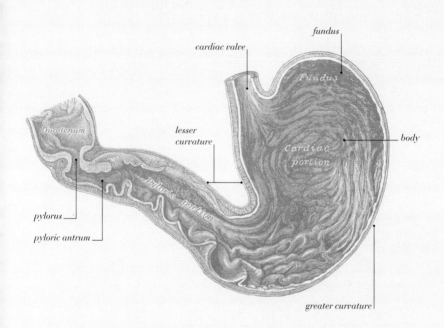

fundus

cardiac valve

Fundus

lesser curvature

body

Duodenum

Cardiac portion

Pyloric portion

pylorus

pyloric antrum

greater curvature

Stomach and top of small intestine • cross-section ◬

When the innermost layer of muscle around the
stomach contracts, it creates complex wrinkles
(called rugae), which churn the food and digestive
fluid inside violently together to assist digestion.
The pylorus (or pyloric valve) at the far end of the
stomach is a sphincter formed by a much tighter
band of muscle than the cardiac valve *(see p. 266)*.
It controls the movement of food into the intestines.

THE SMALL INTESTINE

The small intestine is the longest and narrowest part of the digestive system. Its length varies greatly between individuals, but averages over 20 feet (6 meters). It is described from a medical point of view as being of three parts: the duodenum, jejunum, and ileum (not to be confused with the ilium of the pelvis). The duodenum is the shortest and widest part of the small intestine, traveling in almost a complete circle starting at the pylorus of the stomach *(see p. 271)*. Positioned immediately after the stomach, the contents of the duodenum are naturally acidic, but it is also the point at which the ducts transporting digestive fluids from the pancreas and liver enter the intestine. As these mix with the partially digested food, they continue the process of

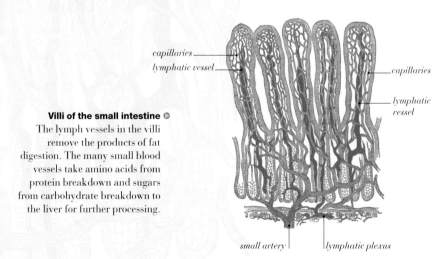

capillaries

lymphatic vessel

capillaries

lymphatic vessel

Villi of the small intestine ◑
The lymph vessels in the villi remove the products of fat digestion. The many small blood vessels take amino acids from protein breakdown and sugars from carbohydrate breakdown to the liver for further processing.

small artery

lymphatic plexus

⊙ **Duodenum within the abdomen**
The duodenum joins on to the
stomach above and the jejunum
below (neither is shown).

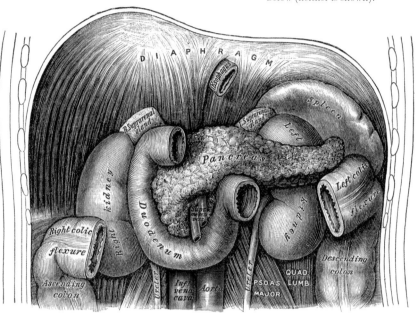

breaking down proteins and carbohydrates, convert the fats to forms that
can be easily absorbed and used by the human body, and raise the pH of the
mixture, making it neutral or slightly alkaline. The two main parts of the small
intestine then transport the mixture slowly along as it digests, and release more
digestive enzymes into the food. These two sections are covered with small,
finger-like projections called villi, which contain large numbers of small blood
and lymph vessels to remove the useful molecules from the food in the intestine
into the body for storage or immediate use.

THE LARGE INTESTINE

The large intestine is about five feet (1.7 meters) long, starting at the bottom of the abdomen and running up one side, across the top, and down the other side before reaching the anus, where waste is expelled from the body. Like the small intestine, it is viewed as having three parts: the cecum, colon, and rectum. By the time food enters the large intestine, digestion is essentially complete—no further digestive enzymes are added here, although the cecum is responsible for absorbing a variety of mineral salts, and has a strong muscular wall to mix its contents thoroughly. The cecum and colon are also responsible for extracting most of the water from their contents, leaving the solid feces that are stored in the rectum before being passed out of the body through the anus.

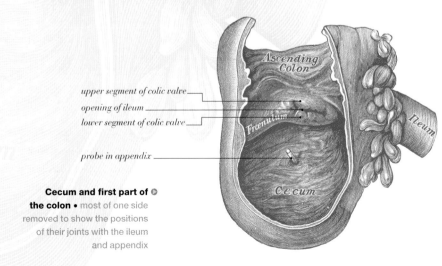

upper segment of colic valve

opening of ileum

lower segment of colic valve

probe in appendix

Cecum and first part of ○
the colon • most of one side
removed to show the positions
of their joints with the ileum
and appendix

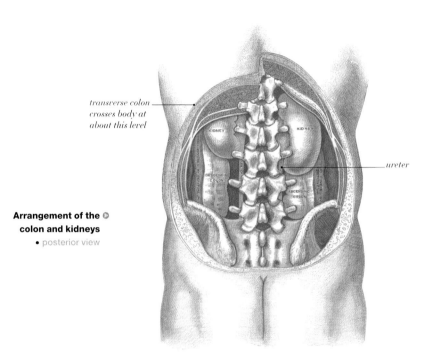

transverse colon
crosses body at
about this level

KIDNEY

KIDNEY

DESCENDING
COLON

ASCENDING
COLON

ureter

Arrangement of the ○
colon and kidneys

• posterior view

The appendix, a much smaller tube opening from the bottom of the cecum, does not appear to serve any particular purpose in the human digestive system, and is generally believed to be the last remnant of a much larger cecum possessed by one of our distant ancestors. Certainly, herbivorous mammals have a proportionally larger cecum and little or no appendix, while carnivores that eat little or no plant matter have little or no cecum, but a proportionally larger appendix. The rectum, the last part of the intestine, absorbs nothing at all from its contents and is merely a storage area—feces that spend a long time in the rectum, however, can be returned to the colon, where more water will be extracted from them.

THE LIVER

The liver performs a large number of different functions for the body, of which the three most important are related to digestion. These are: producing bile, which assists with the breakdown and absorption of fats in the small intestine; completing the breakdown of carbohydrates into glucose (the main sugar the body uses) and glycogen (which is more easily stored); and processing various chemicals removed from the food by the intestines into chemicals that are more directly useful to other parts of the body. In addition to these, it is responsible for removing toxins including caffeine and alcohol from the

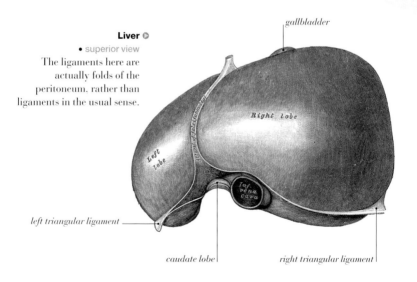

Liver ⊙

• superior view
The ligaments here are actually folds of the peritoneum, rather than ligaments in the usual sense.

gallbladder

Right Lobe

Left Lobe

Inf. vena cava

left triangular ligament

caudate lobe

right triangular ligament

bloodstream and rendering them harmless, storing various vitamins and minerals, and breaking down red blood cells that are no longer able to perform their functions properly. It removes numerous other chemicals, ranging from unwanted hormones (e.g. excess insulin) to waste products of cell metabolism, such as ammonia from the bloodstream. In a fetus it also produces red blood cells, although the bone marrow takes over this task before birth. Although the liver has far greater capacity for self-repair than most other organs, it can still be destroyed, especially by an overload of toxins, and the effects of this are obviously severe.

Liver ◉

• inferior view

The liver is unusual in having two entirely separate supplies of blood. It receives the normal oxygenated blood required to keep the cells alive from the hepatic artery, while the portal vein brings blood from the intestines, so the liver can remove all the useful and harmful chemicals that have been absorbed from the products of digested food.

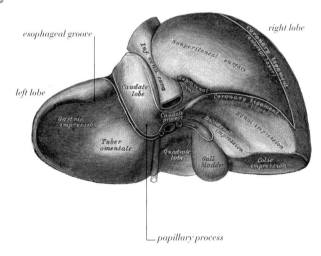

esophageal groove

right lobe (coronary ligament)

Nonperitoneal surface

Inf. vena cava

left lobe

Caudate lobe

semilunar

Coronary ligament (posterior layer)

Caudate process

Renal impression

Gastric impression

Duodenal impression

Tuber omentale

Quadrate lobe

Gall bladder

Colic impression

papillary process

THE KIDNEYS

The kidneys are a pair of bean-shaped organs, positioned roughly level with the bottom of the rib cage at the back of the abdomen. They work to filter the blood in a similar way to the liver and to a lesser extent the spleen, but for different substances. Unlike the liver, the kidneys function solely to remove things permanently from the bloodstream and from the body. In particular they remove a chemical called urea, which is a waste product from protein metabolism. Each kidney consists of millions of tiny filters (nephrons) that pass excess water and unwanted chemicals from the blood into minute tubes (renal tubules). The process is more complicated than this, however: large amounts of water—along with urea and several other chemicals—are removed from the blood by the nephrons. This allows the body to control the concentration of these substances in the bloodstream by changing the amount that is returned from the tubules to the blood. Partway along the tubule, most of the water and varying amounts of some of the chemicals, including sugars, amino acids, and metal ions, are returned to the bloodstream by osmosis. The remainder continues to the bladder to be expelled from the body as urine. The kidneys also produce two hormones. One hormone increases the production of red blood cells in the bone marrow, while the other encourages the deposition of calcium to strengthen bones. If the kidneys fail, it is necessary both to filter the blood (dialysis) and provide an alternate source of these hormones. In the short term, dialysis alone will keep a person alive, but the longterm effects are dangerous without the hormones.

Kidney

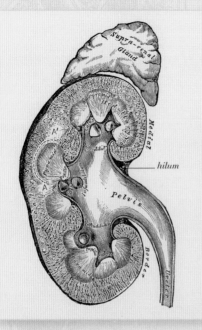

- vertical section (the vertebral column would be to the right)

The pelvis of the kidney is a large open space into which the renal tubules pass their liquid (through structures called the renal or malpighian pyramids), and which forms the open end of the ureter.

Supra-renal Gland

Medulla

hilum

Pelvis

Border

Ureter

Kidneys in position

- posterior view

Each kidney has an indentation on the side closer to the spine, the hilum, through which the renal artery enters and the renal vein and ureter leave the kidney.

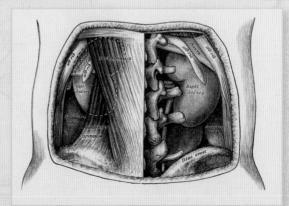

THE URETERS, BLADDER, AND URETHRA

The ureters are the two tubes that run from the kidneys down to the bladder, a distance of about 12 inches (30 cm). These have a muscular coat that tightens and relaxes, similar to peristalsis in the digestive system, to ensure urine moves away from the kidneys (urine flowing backward into the kidneys can lead to serious infections). The distance between the ureters as they enter the bladder changes over the course of the day depending on the amount of liquid in the bladder. The bladder is a hollow, balloon-like organ, lined with a special stretchy layer capable of expanding and contracting as the amount

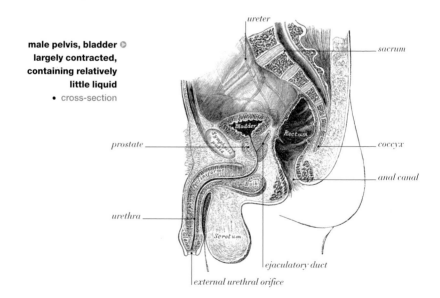

male pelvis, bladder ◐
largely contracted,
containing relatively
little liquid
• cross-section

ureter

sacrum

prostate

coccyx

anal canal

urethra

ejaculatory duct

external urethral orifice

of liquid contained within changes, and surrounded by bands of muscle that contract when urinating. Urine is carried out of the bladder by a tube called the urethra, which is normally held closed where it leaves the bladder by a very tight sphincter muscle to prevent leaks. The signals telling the bladder muscles to contract also relax this sphincter. It is incorrect to say that women's bladders shrink during pregnancy, but the expanding uterus applies a steadily increasing pressure, which has a similar effect.

Male pelvis, ◉
bladder nearly full

• cross-section

The average human bladder has a maximum capacity of about a pint (0.5 liter). A woman's urethra, is only a couple of inches (5 cm) long; the male urethra, which runs to the glans of the penis, is normally around four times this length.

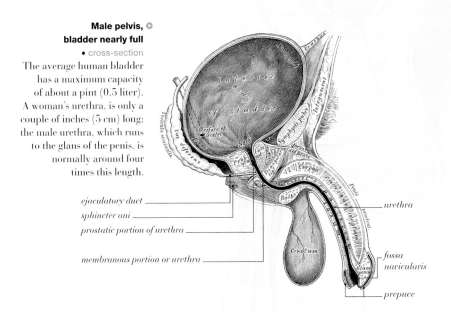

ejaculatory duct

sphincter ani

prostatic portion of urethra

membranous portion or urethra

urethra

fossa navicularis

prepuce

PITUITARY GLAND, PINEAL GLAND, AND HYPOTHALAMUS

Several parts of the endocrine (hormone-producing) system are found within the brain. The pineal gland (or body), situated close to the center of the brain, produces the hormone melatonin, which plays a major part in the control of the body clock. Some prehistoric amphibians had pineal glands that were directly exposed to the light through a hole in the skull, but the human pineal gland is instead connected to a part of the brain called the superchiasmatic nucleus, which receives information from the eyes about light levels and so controls the body clock. The pituitary gland, which is about the size of a pea, sits at the base of the brain and produces a large number of hormones that control many parts of the body. These include a number of the hormone-producing glands farther down the body, but the pituitary gland also has direct effects on growth, blood pressure, metabolic rate, the reproductive organs, and pregnancy. The pituitary gland itself is controlled by the hypothalamus, deeper in the brain, which acts as a link between the nervous and endocrine systems. The hypothalamus receives nerve signals that inform it about the current condition of the body, and releases appropriate neurohormones through a special chemical pathway into the pituitary gland, which then releases the hormones required to make any adjustments.

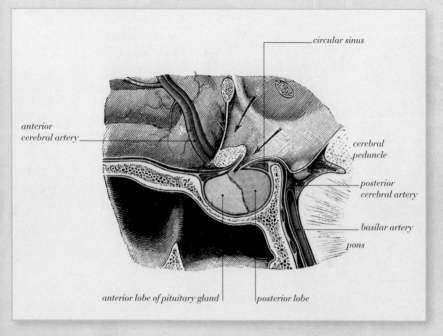

circular sinus

anterior
cerebral artery

cerebral
peduncle

posterior
cerebral artery

basilar artery

pons

anterior lobe of pituitary gland

posterior lobe

○ **Brain and skull**
• vertical cross-section
showing the pituitary gland

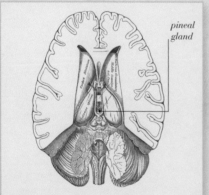

pineal
gland

○ **Brain**
• horizontal cross-section
showing the pineal gland in
the center
The hypothalamus
is an area of the brain
immediately underneath
the thalamus on each side.

THE THYROID AND PARATHYROID GLANDS

The thyroid, a butterfly-shaped gland, lies at the front of the neck, just below the Adam's apple. The parathyroid glands, which can number anywhere from four to eight, are immediately behind it to the sides of the trachea. The thyroid produces several hormones, the most important of which are involved in regulating the metabolism and controlling cell development. In particular, an inadequate quantity of these hormones during childhood (either because the thyroid gland is damaged or absent or because the element iodine, a component of the hormones, is in short supply) can lead to growth problems with the body as a whole and individual organs, including the brain. The parathyroid glands produce only a single hormone, which controls the concentration of calcium in the blood. Many parts of the body—particularly the nervous system—are affected by calcium levels, and will not function properly if there is either too much or too little.

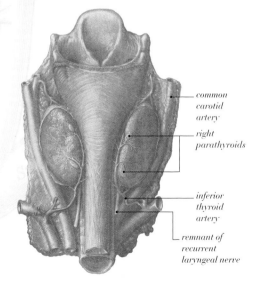

common carotid artery

right parathyroids

inferior thyroid artery

remnant of recurrent laryngeal nerve

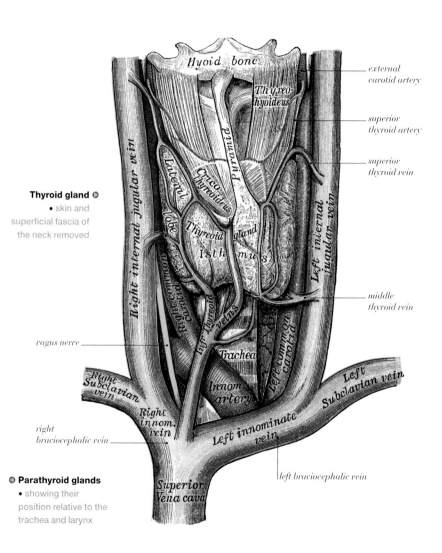

Thyroid gland ⊘
• skin and superficial fascia of the neck removed

⊘ **Parathyroid glands**
• showing their position relative to the trachea and larynx

Hyoid bone

Thyreo-hyoideus

external carotid artery

superior thyroid artery

superior thyroid rein

Patuam

Crico-Thyroideus

Lateral

Lobe

Thyreoid gland

Isthmus

Right internal jugular vein

Left internal jugular vein

middle thyroid rein

Right Thyroid veins

Infr. Thyroid veins

ragus nerve

Trachea

Left common carotid

Right Subclavian vein

Left Subclavian vein

Innom. artery

Right innom. vein

Left innominate vein

right braciocephalic rein

left braciocephalic rein

Superior Vena cava

285

THE THYMUS GLAND

The thymus gland is important in the development of the immune system in children and adolescents, although it remains active throughout life. It grows steadily from before birth until the end of puberty, after which it begins to shrink, and most of the space it previously occupied is eventually used for fat storage. This takes some time, however, with the gland returning to the size it had at birth, which is typically just under half its peak weight of about one-third of an ounce (8 grams), by the time the person is in their fifties. It continues to shrink until death, however, weighing only a few grams in a typical 80-year-old. The thymus converts lymphocytes into several different types of T-cells, which form a major part of the immune system *(see p. 134)*. Some of these move on into the bloodstream, where they are responsible either for controlling the immune response to a threat (loss of these cells is one of the characteristics of AIDS) or for actively hunting down and destroying the infection. The function depends on the T-cell type. Other T-cells remain in the thymus, where they help to educate the next generation of lymphocytes to become T-cells, a process only a small proportion of the cells survive.

Thymus of a full-term baby ◉

• in position immediately behind
the sternum at top of chest
The thymus usually takes a form similar
to this, with two connected but distinct
lobes of lymphatic tissue. Variations
in shape are fairly common including
complete separation of the two lobes or their
combination into a single indistinct lump.

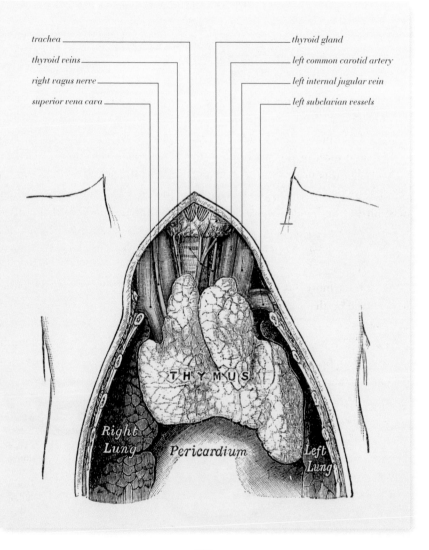

trachea — thyroid gland

thyroid veins — left common carotid artery

right vagus nerve — left internal jugular vein

superior vena cava — left subclavian vessels

THYMUS

Right Lung

Pericardium

Left Lung

THE PANCREAS

The pancreas is a long, pointed organ that runs across the back of the upper abdomen *(see p. 272)*. It performs two very different functions. The first is to produce a large number of different digestive enzymes that are required to break down and absorb a wide variety of fats and proteins. These enzymes are collected by a series of small ducts along the length of the organ, and drain into the duodenum of the small intestine as part of a mixture of chemicals produced by the pancreas. This mixture of chemicals also contains large numbers of bicarbonate ions to neutralize the stomach acid as it enters the intestine with the partly digested food. If the ducts become blocked, this potent

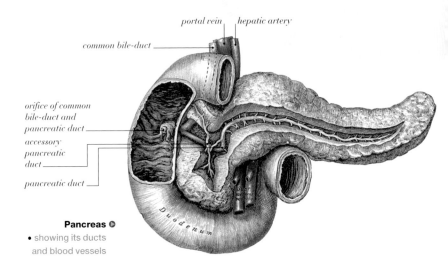

portal vein   *hepatic artery*

common bile-duct

orifice of common bile-duct and pancreatic duct

accessory pancreatic duct

pancreatic duct

Duodenum

Pancreas ◉
• showing its ducts and blood vessels

mixture may become active within the pancreas, which will then start to digest itself. Damage to the pancreas is extremely dangerous for the same reason: a leak of digestive enzymes will soon begin to eat through the surrounding tissue. The second function of the pancreas is to control the amount of sugar in the bloodstream. It does this by producing several hormones that affect the behavior of the liver and other energy stores within the body. Insulin, the most well-known of these hormones, instructs the liver to remove sugar from the bloodstream and store it as glycogen, while others—most notably glucagon—have the opposite effect. Since glucose is necessary for the brain to function properly, hypoglycemia (too little sugar in the bloodstream) is much more dangerous than hyperglycemia (too much sugar in the bloodstream), but the visible symptoms of the two can be very similar.

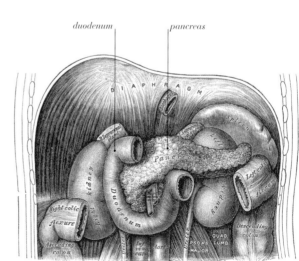

Pancreas and ◉ its immediate surroundings
• with the stomach removed (normally located in front of the pancreas)

THE ADRENAL GLANDS

The adrenal glands, sometimes called the suprarenal glands, are found on either side of the body, on top of the kidneys. Structurally, the adrenal gland is divided into two separate parts, the adrenal medulla inside the gland, and the adrenal cortex, which forms its surface. The two parts manufacture different hormones. The adrenal medulla produces the hormones epinephrine and norepinephrine. These are produced in response to nerve signals from the brain warning that some form of strenuous activity is likely to occur, and have related effects on a number of organs, such as increasing heart rate, blood pressure, oxygen intake, and metabolic rate, which are responsible for the "fight or flight" response. The adrenal cortex produces several different hormones, which are all manufactured from cholesterol, one of many substances that is vital for the body's normal functioning but is dangerous in excessive quantities. Although cholesterol can occur in food, it is also produced by several organs (including the liver and adrenal glands) from other digestion products. These hormones perform several different functions which include controlling the levels of sodium and potassium ions (aldosterone) and increasing the rate of metabolism by encouraging the breakdown of fats and amino acids and the production of extra glucose (cortisol). The adrenal cortex also assists in the production of various steroids including testosterone, a type of hormone that is involved in the growth of muscle and other tissue.

Adrenal glands • posterior view ▾

The two adrenal glands have different shapes: the right adrenal gland has a nearly triangular cross-section, while the left, which is slightly larger, is roughly crescent shaped.

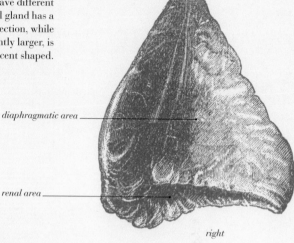

diaphragmatic area

renal area

right

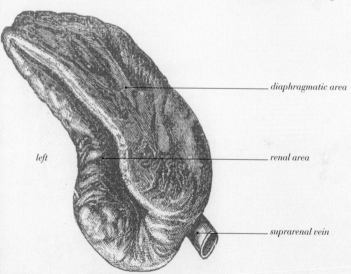

diaphragmatic area

left

renal area

suprarenal vein

THE TESTES

The testes, or testicles, are not attached to any ligaments or muscles, but are instead suspended from the spermatic cord. The spermatic cord contains the blood vessels and nerves, which supply each testicle, and the vas deferens, which carries sperm away. The scrotum, the pouch of skin in which the testes rest, also provides additional support. Their main function is to produce sperm, but they also act as part of the endocrine system *(see p. 282)* as the main producers of various male sex hormones (i.e. those involved in puberty, the sex drive, and sexual activity). The body of the testicle consists of tiny tightly coiled tubes (the seminiferous tubules) which are lined with cells that produce sperm using a special form of cell division called meiosis. The sperm are not immediately able to swim from the seminiferous tubules, however; they move first to the epididymis, a tube in which they mature over a period of two to three weeks before they are ready to be ejaculated.

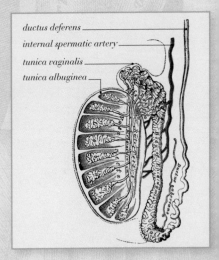

ductus deferens
internal spermatic artery
tunica vaginalis
tunica albuginea

Detailed cross-section through interior of the testicle • showing arrangement of sperm-generating components and ducts that carry sperm to the epididymis and subsequently to the vas deferens

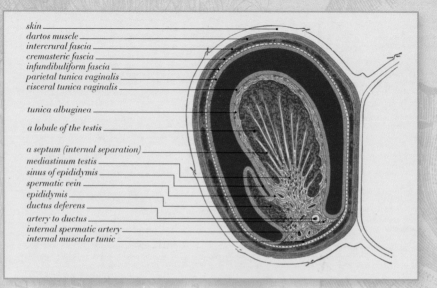

- skin
- dartos muscle
- intercrural fascia
- cremasteric fascia
- infundibuliform fascia
- parietal tunica vaginalis
- visceral tunica vaginalis
- tunica albuginea
- a lobule of the testis
- a septum (internal separation)
- mediastinum testis
- sinus of epididymis
- spermatic vein
- epididymis
- ductus deferens
- artery to ductus
- internal spermatic artery
- internal muscular tunic

Cross-section through the scrotum and left testicle • showing the many protective layers surrounding the sensitive inner parts

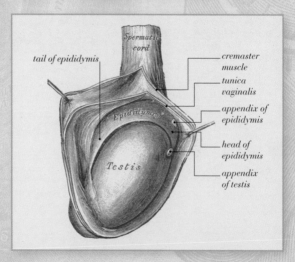

- Spermatic cord
- tail of epididymis
- cremaster muscle
- tunica vaginalis
- Epididymis
- appendix of epididymis
- head of epididymis
- Testis
- appendix of testis

Testicle suspended by its spermatic cord • each testicle is protected by a membrane pouch called the tunica vaginalis, which forms from the peritoneum but becomes separated from it as the testicles descend from the abdomen into the scrotum

THE PROSTATE GLAND AND SEMINAL VESICLES

Sperm are stored in the epididymis until they either become too old to be useful, or until the testes receive hormones indicating sexual arousal. They are then passed into the vas deferens and moved up into the pelvis by a series of contractions and expansions of the muscular wall of the tube, similar to those of the ureters and digestive system. As it runs along the underside of the bladder, the vas deferens is joined by a duct carrying fluid from the seminal vesicles. This seminal fluid contains various chemicals that protect the sperm from harm, as well as nutrients, particularly proteins and sugar, that provide energy to the sperm after ejaculation. The two combined ducts then enter the prostate glands where they join the urethra. The prostate also adds further chemicals to the semen, which is the combination of sperm and seminal fluid. These make the mixture slightly alkaline and provide extra nutrients.

◉ Cross-section through the prostate gland

• showing the urethra, the semicircular shape, and the two ejaculatory ducts immediately beneath it

◉ Underside of the bladder

• showing the vas deferens and seminal vesicle on each side joining together just before their tubes enter the prostate gland The ureters do not, as might appear from the image, pass through the seminal vesicles, but do enter the bladder almost directly underneath them.

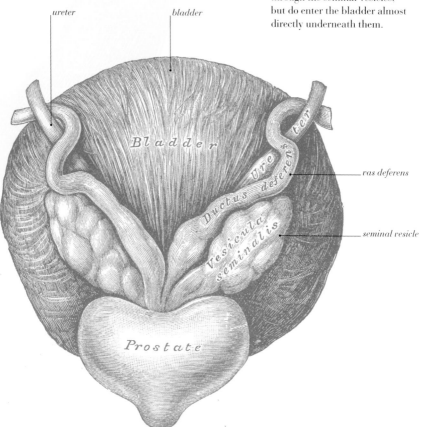

ureter

bladder

vas deferens

seminal vesicle

THE PENIS

The base of the penis is connected to the bones of the pelvis and the area around the perineum by a number of very strong ligaments *(see p. 119)*. The penis itself does not contain any muscles or ligaments. In many animals, the penis is stiffened by a bone, which is retained within a sheath inside the body until the animal becomes aroused, but the human penis does not use this mechanism. Instead, the human penis consists almost entirely of erectile tissue, which has the ability to stretch, holding large quantities of blood. To achieve an erection, muscles around the arteries feeding the penis relax, allowing additional blood to enter it, while muscles surrounding the veins that remove blood from the penis contract, restricting the flow out of the organ.

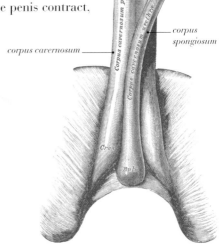

corpus cavernosum penis

corpus cavernosum urethrae

Glans

corpus spongiosum

corpus cavernosum

Crus

Bulb

Base of penis with skin removed ◐
- showing the sections into which the erectile tissue is divided

There are three sections: the corpus spongiosum, a strip that surrounds the urethra along the lower edge of the penis and expands to form the glans (head), at the top end; and the two corpora cavernosa, which run side by side above the corpus spongiosum, forming the main body of the penis.

The penis swells and becomes hard as a result of the pressure. The average length of an erect human penis is between 5 and 6½ inches (13–15 cms). Even at the lower end of that range, however, the human penis is abnormally large relative to the rest of the body when compared with other primates.

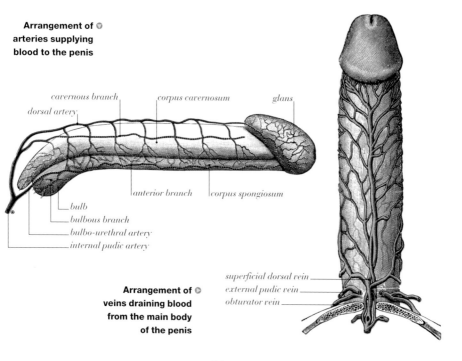

Arrangement of ☉
arteries supplying
blood to the penis

cavernous branch   corpus cavernosum   glans

dorsal artery

anterior branch   corpus spongiosum

bulb
bulbous branch
bulbo-urethral artery
internal pudic artery

Arrangement of ☉
veins draining blood
from the main body
of the penis

superficial dorsal vein
external pudic vein
obturator vein

THE LABIA, CLITORIS, AND VAGINA

The outermost part of the female genitals are the labia, of which there are two pairs. The outer pair are often covered with pubic hair and are relatively thick, with a layer of fatty tissue between their inner and outer surfaces. The inner labia are usually completely covered by the outer pair. These narrow folds of skin surround and help protect the vagina, the urethra, and the tip of the clitoris, which usually has a small hood analogous to the foreskin of the penis. The clitoris, which contains the nerves associated with sexual pleasure, is made of erectile tissue identical to the corpora cavernosa of the penis with which it shares a common origin *(see p. 296)*. Therefore, it also swells up during sexual arousal although, since the exposed part is much smaller, this is less visible. Unlike the penis, most of the clitoris is hidden from view; it runs a short distance up into the body, splitting into two around the vagina. At birth, a fragile membrane called the hymen partially closes the vagina. This is usually broken at the onset of sexual activity or by strenuous non-sexual activities such as horseback riding or gymnastics.

Female genitals ❯

• inferior view, stretched open to make all the external parts visible
The external parts of the female genitals are sometimes collectively called the vulva or pudendum. These terms do not include the vagina, which is the strong muscular tube running from the cervix *(see p. 300)* to the outside of the body.

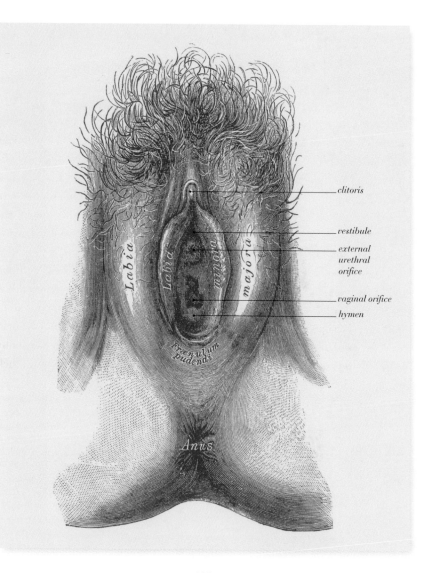

clitoris

vestibule

external
urethral
orifice

vaginal orifice

hymen

THE CERVIX AND UTERUS

The cervix, which forms the neck of the uterus (or womb), is located at the internal end of the vagina. This is normally a very narrow passageway, at an angle of almost 90° to the vagina, but is able to stretch enormously during childbirth to allow the baby to pass through. It also pulses during orgasm, in a way that is believed to help semen move from the vagina into the uterus, and so increase the chance of pregnancy. The main part of the uterus consists of a very strong body of muscle containing a large number of blood vessels. During pregnancy, the layers of muscle stretch and slide across one another to cope with the expansion of the cavity inside the womb. Normally this cavity is very small, and roughly pyramidal in shape, and the expansion is relatively slow during the initial stages of pregnancy. Subsequently, the steadily increasing size of the fetus and of its support structures, the placenta and amniotic sac, force the walls of the uterus outward, causing the familiar shape of pregnancy, and applying steadily increasing pressure to the other abdominal organs.

Female pelvis ○

• cross-section showing the vagina and uterus in their normal contracted condition
In this state, the length of the vagina is about 4 inches (10 cm), while that of the uterus is typically around 3 inches (7.5 cm). The size of the uterus changes continuously during pregnancy. The vagina stays much the same size until labor, when it shortens and widens.

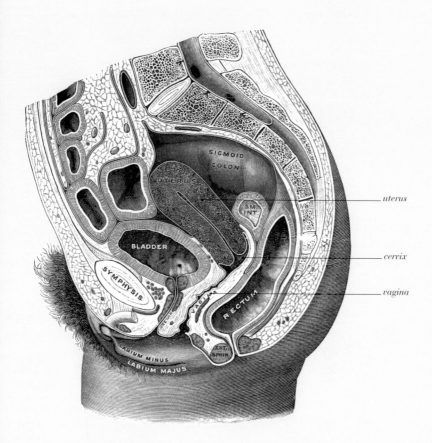

SIGMOID
COLON

UTERUS

SM.
INT.

uterus

BLADDER

cervix

SYMPHYSIS

vagina

VAGINA

RECTUM

LABIUM MINUS

LABIUM MAJUS

EXT.
SPHIN.

THE OVARIES AND FALLOPIAN TUBES

Unlike the testicles, which manufacture sperm continuously from puberty onward, the ovaries of a newborn girl already contain every egg (ovum) her body will ever manufacture. Each of these is stored in a special small capsule called a graafian follicle, and these normally mature and release the egg they contain at a rate of about one per month, halfway between menstrual periods, from puberty onward. These eggs are then pushed down the fallopian tubes toward the uterus, over a period of several days, by the waving movements of small, finger-like structures lining the tube. If the egg is fertilized by a sperm during this time, it will then normally implant itself on the wall of the uterus; this causes a number of hormonal changes that mark the beginning of pregnancy. Like the testicles, the ovaries are the body's major producer of sex hormones from puberty onward.

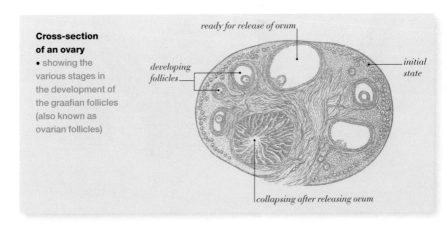

Cross-section of an ovary
• showing the various stages in the development of the graafian follicles (also known as ovarian follicles)

ready for release of ovum

developing follicles

initial state

collapsing after releasing ovum

Uterus, fallopian tube, and ovary, and the arteries that supply them

• posterior view

An egg released from an ovary must cross a small open space of about 1/2 inch/1 cm before reaching the end of the fallopian tube, down which it will move to the uterus. At the end close to the ovary, the fallopian tube opens into the space between the two layers of the peritoneum. While in a man this cavity is sealed, in a woman it is actually open to the outside through the vagina.

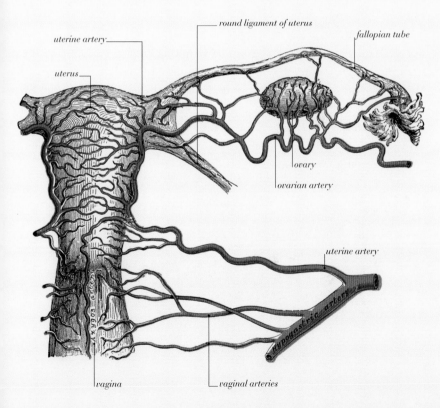

THE BREASTS AND LACTATION

Both men and women possess mammary glands in the chest behind the nipples, and before puberty the structures are exactly the same in both sexes. At puberty, the female sex hormones cause growth of the mammary gland, and changes in the surrounding areas, forming the breasts. (If a man receives large doses of these hormones—for example, as a medical treatment—the effects are exactly the same, and the resulting breasts are fully functional and able to produce milk under suitable circumstances.) One unusual feature of the area around the mammary gland is the fatty layer of the superficial fascia, which is thicker here than in most parts of the body. It divides in two, passing both in front of and behind the part of the breast responsible for producing milk. After the first few days of breast feeding, when the initial milk contains antibodies to assist the newborn's immune system, enough milk is produced to match the demand from the infant. This will normally continue for as long as breast feeding happens regularly.

**Partially dissected ◉
human breast**

The nipple, which can be pulled erect by the small amounts of muscle around it, as a result of either sexual arousal or cold, is directly over a small hole in the fatty layer through which pass the milk-carrying (lactiferous) ducts. There are normally around 20 of these in each breast, each of which has a small reservoir called the ampulla for storage of milk immediately beneath the fatty layer as they spread out from the nipple. Beyond these, the milk ducts branch repeatedly to collect milk from the numerous small hollow lobes that make up the mammary gland proper.

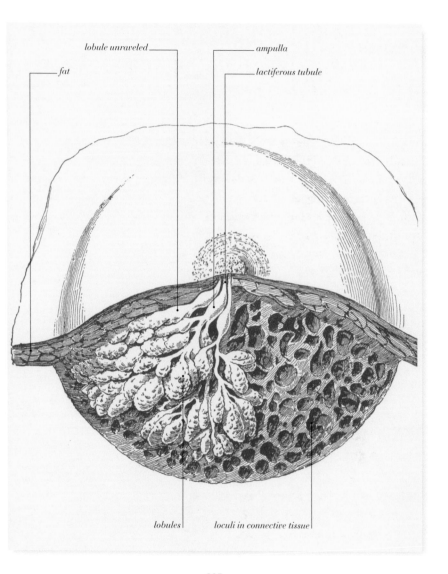

fat

lobule unraveled

ampulla

lactiferous tubule

lobules

loculi in connective tissue

OVUM AND SPERM

Sexual reproduction requires special cells called gametes; the egg and sperm are the female and male gametes respectively. The defining difference between gametes and normal cells is that gametes contain only half the normal number of chromosomes, which means they have only half the normal amount of genetic material. Chromosomes exist in pairs, and normal cell division (called mitosis) reproduces all chromosomes in both of the daughter cells. Gametes are produced by a special type of cell division called meiosis, in which each

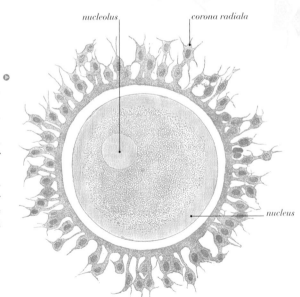

nucleolus

corona radiala

Human egg ○
A human egg is the largest type of individual human cell, just large enough to be visible to the naked eye, but it is still extremely small: less than $1/100$th of an inch (0.25 mm) across.

The outermost layer is designed to be difficult for sperm to cross, to reduce the chance of a damaged sperm fertilizing the egg.

nucleus

daughter cell receives only one chromosome out of each pair. In addition to this, some genetic material is swapped randomly between the chromosomes within each pair to increase the genetic variation in the gametes, and thus in the offspring produced. When the egg is fertilized by a sperm, the individual chromosomes from the two gametes pair up, producing a cell with the full normal complement of chromosomes.

Human sperm ○
• surface view (A); profile view (B); head, neck, and connecting piece, highly magnified (C)
The sperm is much smaller than an egg and consists of a relatively sharp, pointed head and a long tail with which the sperm swims toward its destination. The perforator penetrates the outermost layer of the egg and delivers the genetic material.

GLOSSARY

POSITIONS AND VIEWS

Anatomists divide the upright
human body into anatomical
planes—imaginary vertical or
horizontal lines. When discussing
the anatomy of different body
parts and when referring to
anatomical illustrations,
specialist terms such as those
listed below are used to describe
position, or the direction of view.
In common with other branches
of science, the majority of terms
are derived from Latin.

AFFERENT Latin *ad* (to) and
ferre (to bring), and meaning
"bringing to." Usually used to
refer to nerve axons which carry
data from a sense organ to where
it is to be processed.

ANTERIOR Latin *ante*, meaning
"before," or "in front."

AXIAL position in relation to the
central nervous system: "intra-
axial" is within the CNS (central
nervous system), and "extra-
axial" is outside the CNS.

CONTRALATERAL Latin *contra*
(against or opposite) and *latus*
(side), and meaning "the
opposite side."

CROSS-SECTION a transverse cut
through a structure or tissue. The
opposite of a cross-section is a
longitudinal section.

DEEP position of a structure,
denoting its remoteness from the
skin or outer surface.

DISTAL Latin *distare* (to stand
apart), and referring to those
parts of the anatomy farthest
from the body's center, or
farthest from a limb's point
of joining the trunk.

DORSAL Latin *dorsum*, meaning
"back." *See* posterior.

EFFERENT Latin *ex* (from) and
ferre (to bring), and meaning
"bringing away."

INFERIOR (as prefix, infra)
Latin *inferus* (below), and
meaning "lower" or
"from below."

IPSILATERAL Latin *ipsi* (self)
and *latus* (side), and meaning
"the same side."

LATERAL Latin *latus* (side), and
meaning "at the side" or "from
the side."

MEDIAL Latin *medius* (middle),
and meaning "at the middle" or
"from the middle."

PALMAR Latin *palma* (palm),
and referring to the palm of
the hand.

PLANTAR Latin *planta* (sole),
and referring to the under-
surface (sole) of the foot.

POSTERIOR Latin *post* (behind),
and meaning "behind" or "from
behind." *See* dorsal.

PROXIMAL Latin *proximus*
(nearest), and refers to those
parts of the anatomy nearest to
the body's center, or nearest to a
limb's point of joining the trunk.

SAGITTAL Latin *sagitta* (arrow),
referring to the arrow-shaped
suture dividing the left and right
parietal bones of the skull, and
by extension a view or section
taken parallel to this suture.

SUPERFICIAL position of a
structure denoting its nearness
to the skin or outer surface.

SUPERIOR (as prefix, supra-)
Latin *super* (above), and
meaning "above" or
"from above."

TRANSVERSE Latin *trans*
(across) and *vertere* (to turn),
and referring to a view or section
that crosses the long axis (an
imaginary line through the center
of the structure being studied,
from end to end).

VENTRAL Latin *venter* (belly),
and referring to a view, or a
surgical approach, from the
belly side. *See* anterior.

MOVEMENT

These anatomical terms describe the principal movement of parts of the anatomy, in particular the limbs, joints, and muscles. The names of individual body structures, such as muscles and joints, are generally derived from these basic terms.

ABDUCTION Latin *ab* (from) and *ducere* (to lead), and meaning "leading away" or "moving away." Abductor muscles move a limb away from the center-line of the body.

ADDUCTION Latin *ad* (to) and *ducere* (to lead), and meaning "leading toward" or "moving toward." Adductor muscles move a limb toward the center-line of the body.

ANTEFLEXION Latin *ante* (before) and *flectere* (to bend), and referring to an anatomical structure (usually soft-tissue) that bends forward.

ANTEVERSION Latin *ante* (before) and *vertere* (to turn), and referring to an anatomical structure (usually bony) that turns forward.

APPOSITION Latin *appositus* (side by side), and referring to structures that are in contact with each other.

ARRECTOR Latin *arrigo*, meaning "to raise," or "that which raises." The arrector pili is a muscle that causes an individual hair to stand on end.

ARTICULATION Latin *artus* (a joint), and referring to structures that work together as a joint.

DEPRESS Latin *depressus*, meaning "to press down." A depressor muscle is one that draws down a body part.

DILATE Latin *dilatare* (to spread wide), and meaning "to distend." A dilator muscle is one that dilates a body part.

EVERSION Latin prefix *e* (ex) (from) and *vertere* (to turn), and meaning "turning outward" or "inside out."

EXTENSION Latin *ex* (out) and *tendere* (to stretch), and referring to the process of straightening or extending. Extensor muscles extend or straighten out parts of the body.

FIXATION Latin *figere*, *fixus* (to fasten, fastened), and referring to a structure that holds something in place.

FLEXION Latin *flectere* (to bend), and referring to the act of bending. Flexor muscles cause a body part to bend.

INVERSION Latin *in* (in) and *retere* (to turn), and meaning "turning inward," "inside out," or "upside down."

LEVATE Latin *levare*, meaning "to raise." A levator muscle is one that raises a body part.

OCCLUSION Latin *occludere* (to close or shut), and referring to the closing or blocking of a hollow organ, or to the apposition of upper and lower teeth when the jaws are closed.

RETROFLEXION Latin *retro* (backward) and *flectere* (to bend), and referring to an anatomical structure (usually soft-tissue) that bends backward.

RETROVERSION Latin *retro* (backward) and *vertere* (to turn), and referring to an anatomical structure (either soft-tissue or bony) that turns backward.

ROTATION Latin *rotare* (to turn), and referring to the act of turning.

TENSION Latin *tendere* (to stretch), and referring to structures that are stretched or tightened.

TORSION Latin *torquare* (to twist), and referring to structures that are twisted.

GLOSSARY

COMMON ANATOMICAL TERMS

This glossary covers common anatomical terms used in this book and in other sources of reference. The terms describe anatomical structures and functions which are found in humans and in many animals. The definitions here, however, relate only to human anatomy.

ACOUSTIC refers to anatomical structures related to sound and the sense of hearing. *See also* auditory.

ADENOID describes a structure that has the shape of a gland; especially, a mass of enlarged lymphatic tissue at the back of the throat.

ADHESION tissue which adheres, or becomes stuck to, other tissue, which often takes place as a result of inflammation or surgery.

ADIPOSE fatty tissue used to store food and energy, and as insulation for the body and some of its organs.

ADRENAL near to the kidney; especially the adrenal glands, two ductless glands situated above the kidneys which secrete hormones including corticosteroids.

APONEUROSIS connective tissue in the form of flat sheets of tendon which attach muscle to bone, and which form the deep fascia overlying the muscles.

ARTERY a blood vessel, part of the circulatory system, through which oxygen-rich blood is pumped by the heart.

ATROPHY wasting which occurs when tissue is deprived of nourishment.

AUDITORY structures to do with hearing and the ear.

AUTONOMIC functioning without direction. The autonomic nervous system is responsible for those functions which occur involuntarily, such as the beating of the heart.

BASILAR relating to the base of a pyramid or broad structure.

BIFID split or forked; structures which are divided into two parts.

BILATERAL occurring on both sides of a structure.

BUCCAL describes structures and tissue related to the cheek.

BURSA a sac containing fluid, used as a cushion between structures.

CAPILLARY a very narrow and delicate blood vessel.

CARDIAC describes structures generally related to the heart, either by function or proximity.

CARTILAGE firm but flexible connective tissue that forms part of the joints between bones, or that provides support to softer tissues in, for example, the nose and the ear.

CAUDAL Latin *cauda* (tail), refers to structures that are tail-like or at the lower end of the body.

CERVICAL Latin *cervix* (neck); for example, cervical vertebrae are bones that form the upper part of the vertebral column in the neck, while in females the cervix is the "neck" of the uterus.

CONDYLE Greek *kondulos* (knuckle); a rounded process at the end of a bone which articulates, or forms part of a joint, with another bone.

CONNECTIVE TISSUE the strong, fibrous tissue that forms tendon and cartilage; also the loose mesh of tissue that locates and supports internal organs.

CORONARY Latin *corona* (crown), describing those structures which encircle, and particularly the coronary arteries which encircle the heart.

CORPUSCLES minute particles or bodies; especially the red and white "cells" of the blood (although, since the blood's red corpuscles do not have nuclei, they are not really cells).

CORTEX the outer part of an organ; e.g., the renal cortex is the external part of the kidney.

CRUCIATE Latin *crux* (cross); describes structures formed in the shape of the letter x.

DISSECTION Latin *dis* (apart) and *secare* (to cut); describes the process by which anatomical structures are separated from each other, either as part of a surgical procedure, or more generally to reduce a structure to its component parts for examination.

ENDOCRINE describes those glands which secrete a substance directly into the bloodstream.

ENDOTHELIUM the internal lining of the blood vessels, the heart, and the lymphatic vessels.

EPITHELIUM the tissue which forms the internal and external surface of the body, including the skin, the mucous membrane inside the mouth, the endothelium, and the lining of the alimentary canal between mouth and anus.

FACET a small, smooth, bony surface, often the point at which a tendon is attached to the bone.

FASCIA the superficial fascia, composed mainly of adipose tissue, lies beneath the skin and both protects and insulates the structures beneath; the deep fascia, formed of aponeuroses, lies between the superficial fascia and the body's internal structures such as muscle and organs. Some muscles and muscle groups are also sheathed in fibrous fascia.

FIBER a thread or filament; both nerves and connective tissue may be formed of fibers.

FISSURE a cleft (*see* bifid); for example, the cleft between the convolutions on the surface of the brain.

FONTANELLE one of several areas between the various bony parts of the infant skull, formed of a membrane of connective tissue. The fontanelles usually close within two years after birth.

FOSSA Latin for "ditch"; a depression in the surface of a bone (for example, the temporal fossa on the side of the skull) or soft-tissue structure.

GANGLION a group of nerve cells usually contained in a swelling or a knot.

GLAND one of several different secreting organs, which may be ducted or ductless (*see* endocrine); for example, lacrimal glands secrete tears via the lacrimal duct, while adrenal glands secrete hormones directly into the bloodstream.

GLOSSAL relating to the tongue.

GUSTATORY to do with the sense of taste.

HEMISPHERE Greek *hemi* (half) and *sphaira* (ball), and meaning one half of a ball-shaped structure; for example, the brain is formed of two (cerebral) hemispheres.

HEPATIC to do with the liver.

HOMOLOGOUS structures which are homologous share a very similar form but different functions; for example, the female ovary is homologous with the male testicle.

HORMONE a substance produced, usually by a gland, to control the body's function; for example, epinephrene is secreted by the adrenal glands to initiate the "fight-or-flight" reaction in the case of perceived danger.

INNERVATE to supply with nerves.

GLOSSARY

INNOMINATE Latin for "without a name"; the bones of the hip which individually are unnamed are collectively known as the right and left innominate bones.

JOINT the point at which two bones or cartilages meet and around which movement is possible.

LABIAL relating to the lips, or to the external folds of the female genitals.

LIGAMENT a band of fibrous connective tissue connecting bones with each other.

LINGUAL relates to the tongue (*see also* glossal).

LUMBAR the section of the back between the ribs and the pelvis; there are five lumbar vertebrae in the vertebral column.

LYMPHATIC describes the system through which lymph is distributed through the body; although complementary to the cardiovascular system which distributes blood, it is different in that it is not pumped by the heart but is instead propelled by the incidental movements of muscles and surrounding tissues.

MATRIX the structure within which cells are embedded, and in some cases from which new structures are generated; for example, the base of the nail (the nail bed) is a matrix.

MEDIAN Latin *medianus* (the middle), and thus something which is sited at the center of a structure.

MEMBRANE a thin sheet of tissue.

METABOLISM the process by which various organs of the body convert nutrients and oxygen into energy.

MUCOUS MEMBRANE a membrane which secretes mucus.

MUCUS a slippery substance which lubricates and protects the tissues which it covers.

MUSCLE a collection of bundles of fibrous tissue, contained within a sheath of connective tissue, whose contraction or relaxation causes the movement of an anatomical structure.

NASAL relates to the nose.

NODE a collection of cells of similar type, especially lymph nodes; a break or junction in a nerve sheath.

OCULAR refers to the eye, or the sense of sight. (*See* ophthalmic *and* optic.)

-OID Greek *eidos* (shape), referring to structures whose shape resembles that of a well-known object; for example, "ovoid" (eggshaped) and "sigmoid" (curved like the letter *s* or *c*, from the Greek "sigma").

OLFACTORY to do with the sense of smell.

OPHTHALMIC refers to the eye.

OPTIC to do with the sense of sight.

PARASYMPATHETIC describes part of the autonomic nervous system that controls the glands, blood vessels, and internal organs. (*See* sympathetic.)

PECTORAL describes structures at the front of the chest.

PERI- Greek, meaning "round," or "about"; used as a prefix in terms such as pericardium (the membrane which surrounds the heart).

PLEXUS a network, usually referring to one of several areas in the body at which nerves are concentrated; for example, the solar plexus is located behind the stomach, while the cardiac plexus is close to the heart.

PROCESS in anatomy, an outgrowth, usually of bone, and frequently used as the points at which muscles are attached.

PULMONARY to do with the lungs.

RETRACT Latin *re* (again) and *trahere* (to draw), meaning "to pull back" or "to withdraw."

RETRO- Latin for "backward"; used as a prefix in, for example, "retrosternal" (behind the breastbone) and "retroverted" (turned backward).

SECTION in anatomy, a cut surface.

SINUS a hollow cavity within a structure which might contain a gas (such as air), or a liquid (such as venous blood, or lymph); a passageway from the surface of a structure to a deep abscess.

SOMATIC Greek *soma* (the body), and referring to the frame of the body rather than to its internal contents.

SPHINCTER a muscle which is circular in form, and which as it opens and closes acts as a valve; for example, the pyloric sphincter at the distal end of the stomach, between the stomach and the duodenum.

SPINAL CORD part of the central nervous system, the spinal cord occupies the hollow space within the vertebral column between the base of the brain and the lumbar vertebrae. It contains nerves which carry sensory information to the brain, and which control movement; at its center is a tube containing cerebrospinal fluid.

SQUAMOUS Latin *squama* (scale), describing tissue which is scaly; e.g., the outer surface of the skin.

STRIA Latin for a stripe, meaning linear in form; for example, muscles made from fibers laid side-by-side are frequently described as striated.

SULCUS Latin for a groove; especially a groove on the surface of the brain.

SUTURE Latin *suere* (to sew); in anatomy, the seam between two bones, especially in the skull. In the fetal skull the bones of the cranium are separate to allow it to distort as the head passes through the vagina at birth, and following birth these bones fuse along the sutures.

SYMPATHETIC refers to part of the autonomic nervous system that controls glands, blood vessels, and internal organs, and which works in parallel with the parasympathetic system.

TENDON Latin *tendere* (to stretch), referring to the fibrous connective tissues which connect muscle to bone.

THORACIC to do with the thorax (chest).

THORAX the part of the trunk between the neck and the abdomen.

TUBERCLE a small lump or protuberance, usually part of a bone.

VALVE a structure which, when moved, opens and/or closes an aperture or vessel. The chambers of the heart are separated by valves which allow blood to be pumped around the cardiovascular system; and most veins contain valves which prevent gravity, causing the blood contained within the veins from pooling in the legs and feet.

VEIN a blood vessel, part of the circulatory system, through which blood carrying carbon dioxide (a waste product of the body's metabolism) is pumped by the heart toward the lungs.

INDEX

ACKNOWLEDGMENTS

THE illustrations in this book are reproduced directly from *Anatomy of the Human Body* by Henry Gray, F. R. S (1918) and *Anatomy Descriptive and Surgical* by Henry Gray, F. R. S (1905).